C0-CFG-087

PROGRAMMING FOR NUMERICAL CONTROL MACHINES

Second Edition

ARTHUR D. ROBERTS
Formerly Department Chairman, Manufacturing Engineering Technology,
Norwalk State Technical College

RICHARD C. PRENTICE
Manufacturing Engineer,
Sikorsky Aircraft Division of United Technologies

McGRAW-HILL BOOK COMPANY · GREGG DIVISION

New York   St. Louis   Dallas   San Francisco   Auckland   Bogotá   Düsseldorf
Johannesburg   London   Madrid   Mexico   Montreal   New Delhi   Panama
Paris   São Paulo   Singapore   Sydney   Tokyo   Toronto

Library of Congress Cataloging in Publication Data

Roberts, Arthur D
    Programming for numerical control machines.

    1.   Machine-tools—Numerical control.
I.   Prentice, Richard C., joint author.
II.   Title.
TJ1189.R6   1978   621.9′02   77-10982
ISBN 0-07-053156-0

*The editor for this book was Don Hepler, the art supervisor was George T. Resch,
and the production supervisor was May E. Konopka. Cover design by Jackie
Merri Meyer. It was set in Times Roman by Progressive Typographers.
Printed and bound by R. R. Donnelley & Sons Incorporated.*

CONTENTS

PREFACE

In today's world practically all large manufacturing companies own several numerical control (N/C) machines. The smaller plants, too, are finding that tape-controlled machines can make dramatic savings possible. It is estimated that by 1976 there were as many as 40,000 numerical control machines in use in industry, and the rate of increase seems to be limited only by the rate at which these machines can be manufactured. To speed up production, N/C machines are now frequently producing parts for N/C machines. Thus it seems important that anyone who is training for work in industry should have a basic knowledge of what numerical control is, how the tapes are made, and what information they contain.

"Programming" for tape-controlled machines means writing a "manuscript" which contains the proper series of numbers, symbols, and letters so that, when these are punched into a tape, and the tape is "read" by the machine, the machine will perform a series of operations on the workpiece, in agreement with the part drawing.

This book is concerned principally with the actual detailed description of how to do manual programming, commonly referred to as hand programming. Hand programming is basic to the writing of all N/C programs, just as arithmetic is basic to all mathematics. Many tape control programs are still being efficiently handprogrammed, since a great deal of the machining, bending, punching, etc., done on numerical control machines involves relatively simple arithmetical and geometrical computations.

It is not too difficult a step from manual programming to the use of the simpler programs such as AUTOSPOT and AUTOMAP. Then, with a somewhat broader concept of mathematics, the student can learn the APT language, which can be used to program almost any imaginable shape. Chapter 12 in this book gives examples of how N/C programs are written in computer language.

The first four chapters are designed to give the students an overall picture of the history of numerical control, the wide variety of the tape-controlled equipment available, and the basic ideas of the coding systems, axis notation, tapes, and tape-punching equipment.

With this information to build on, Chapter 5 starts with the simplest types of program, and each succeeding chapter adds a new, slightly more complex phase of N/C programming. Thus an instructor may limit the course to the simple basic TAB SEQUENTIAL and WORD ADDRESS concepts or continue with the somewhat more advanced applications of these systems.

The language used is clear, and each chapter on programming begins with a discussion of the unique features and coding system of the type of N/C machine being discussed. Following this is an actual N/C program, with detailed explanations of each step, so that students may learn the material without constant assistance from the instructor.

The number of points or surfaces to be located on the sample problems has purposely been kept small. Thus the basic principles are emphasized and the use of time-consuming, repetitive calculations is avoided. The examples given are designed to emphasize important, practical features of the various types of programs. The method of dimensioning the problems is deliberately varied to emphasize that industry still uses many methods.

New words are clearly defined in the text as they are used. An extensive glossary in the Appendix includes other words and abbreviations which are used in the numerical control industry.

The manuscript forms, and the symbols used, are the same as, or very similar to, those used in industry today. All notations (unless otherwise defined) are according to the recommendations of the Electronic Industries Association (EIA) and the Aerospace Industries Association (AIA). Thus the methods shown in this book will be applicable to most of the N/C machines in school shops, and they will also provide the students with a training that can be of immediate value when they enter industry. Although no two makes of numerical control machines are programmed exactly the same, most manufacturers follow the EIA and AIA standards quite closely.

It is, of course, preferable to have at least a simple numerical control machine and tape-punching equipment available in the school. Then the students can actually experience the ''debugging'' of their own tapes. However, the factors involved in N/C programming can be learned just as thoroughly, even when a school does not have any tape-controlled equipment. One or two visits to nearby plants having numerical control equipment would be valuable.

Details of tooling and fixtures have purposely been omitted. This is a study in itself, and probably no two tool designers would offer the same solution to a given problem. The authors believe that it is wiser for the classes to concentrate on N/C programming techniques and not get sidetracked into discussion on tool design. However, the tooling up for some of the problems can be presented to a separate class in tool design, making an interesting tie-in if the instructor so desires.

The feeds, speeds, and cutting tools used in the examples throughout the book are based on good machining practices. In several instances alternative methods are mentioned, and practical suggestions are included. Special features such as ZERO OFFSET and types of tape readers are in the Appendix, and may be used as time and interest dictate.

Having, in their own teaching, felt the need for a practical "how-to" book on numerical control programming, the authors believe that many others have a similar need. We trust this book will serve the purpose of making it easier for students and instructors to learn more about this very vital subject.

The Second (Metric) Edition

The basic methods of numerical control programming have not greatly changed since 1968. However, N/C machines have become more versatile and some previous options are now practically standard on all but the simplest N/C equipment. The outstanding example is the fact that linear and circular interpolation are now supplied as standard or as low-cost options on practically all N/C machines. The number of models of "machining centers" with automatic tool changes has also increased. The sales of N/C lathes of several designs has increased dramatically since 1968, and N/C punching machines are more widely used, so programming for these is added to the revised edition.

N/C systems are now all solid-state, and CNC (Computer Numerical Control) in a modified form is becoming a valuable, not-too-expensive option, not eliminating the tape but allowing changes to be easily made.

Thus, throughout this revised edition, many minor and major changes have been made. The many "special" abilities of the various N/C machines have not been included, because this is still a book of *basic* programming skills. Once these basics are mastered, the added capabilities can be learned very quickly from a company's programming manuals.

Explanations and drawings have been revised for clarity in several chapters of *Programming for Numerical Control Machines,* the appendices have been updated, and metric taps and tap drills are now shown.

The Metric System Used in this Edition

The greatest change in the second edition is metrication. The United States is officially "going metric." Thus our designs, and therefore our N/C programs, will increasingly be done using metric measurements.

The use of the metric system does not change any of the machining practices, tolerances and fits, or speeds and feeds—it only gives them new numbers, which will, for a short while, seem strange.

As in the past all dimensions on machining drawings were in inches, so in the future all machining dimensions will be in millimetres (mm).

Thus, instead of a 48-in.-long lathe bed, it will be a 1200 mm (rounded figure) bed. Instead of the 15-ft travel of a large planer, it will be 4600 mm travel (or 4.6 m).

Tolerances will usually be carried to one **less** decimal place in millimeters than in inches. The common ±0.001 in. becomes ±0.02 mm (approximately 0.0008 in.). Thus 0.01 mm is close to our "half-thousandth" and a tenth (0.0001 in.) is 0.0025 mm, preferably 0.002 mm (0.00008 in.). Note that the ISO specifies that a zero must precede the decimal point when using millimetre dimensions of less than 1.00 mm.

In revising the dimensioning of the drawings, rounded metric dimensions and tolerances have been used, as might be done when making an original design in metric.

Metric N/C Machines

Many N/C machines now have "switchable inch/metric" controls. The tapes have the same number of digits, but the decimal point has usually been moved one place to the right. Thus an N/C machine having 99.999 in. available on the tape now has 999.99 mm [39.37 in.] capacity—which is sufficient for many machine tables. More digits are used on either or both sides of the decimal point on some N/C machines. The conversion to millimetres is often done "electronically," as the lead screws still have a 0.200- or 0.250-in. pitch.

Acknowledgments

As was true with the first edition, the manufacturers of the N/C machines have been a vital factor by providing new illustrations where needed, the newest programming manuals, and much information. Many individuals helped in the preparation of the first edition by using some of the chapters in classroom teaching and with reviews and suggestions. We wish to acknowledge especially the help of Robert Piccuillo of Cynmar Consultants, Samuel Lapidge and Carl German of Norwalk State Technical College, and George McLoughlin and Warren Anderson of Pitney Bowes Company. And once again Margaret Roberts has typed the many letters and new pages needed for this revised edition. Without these peoples' assistance, the book and this revision could not have been written. Our sincere thanks to them all.

Arthur D. Roberts
Richard C. Prentice

Chapter 1

NUMERICAL CONTROL— PAST AND PRESENT

Today thousands of numerical control machines are in use. Some are in large corporations, and many are in small companies or job shops. The cost of each machine may be under $15 000 or over $2 000 000. They may be used for accurately drilling a few holes in a simple part, or for contour milling a complex shape which is practically impossible to machine by conventional methods. Even though the amazingly rapid growth of N/C machines did not start until about 1960, the achievement was the result of work done by many people, starting in 1947.

1947 was the year in which numerical control was born. It began because of an urgent need. John C. Parsons, of the Parsons Corporation, Traverse City, Michigan, a manufacturer of helicopter rotor blades, could not make his templates fast enough. So, in 1947, he invented a way of coupling computer equipment with a jig borer. Mr. Parsons used punched cards to operate his Digitron system.

1949 was the year of another "urgent need." The U.S. Air Materiel Command realized that parts for its planes and missiles were becoming more complex. Also, as the designs were constantly being improved, changes in the drawings were frequently made. Thus, in their search for methods of speeding production, an Air Force study contract was awarded to the Parsons Corporation. The Servomechanisms Laboratory of the Massachusetts Institute of Technology was the subcontractor.

In **1951** the Massachusetts Institute of Technology took over the complete job, and in **1952**, the prototype of today's N/C machine, a much modified Cincinnati Hydrotel milling machine, was successfully demonstrated. The term **numerical control** was coined at M.I.T.

1

In **1955** about seven companies had tape-controlled machines exhibited at the Machine Tool Show. Most of these were different types of contour-milling machines. Many of these early machines cost several hundred thousands of dollars, and some required trained mathematicians and computers to make the tapes for their complex work. However, the machine tool manufacturers soon realized that N/C was an idea which could also be used in many simpler ways.

In **1960**, according to the *American Machinist,** there were one hundred numerical control machines at the Machine Tool Show in Chicago. Most significant was the fact that the large majority of these machines were for relatively simple point-to-point applications. Many of these machines sold at $50 000 or less, which at this time was quite inexpensive for tape-controlled machines. Sales of N/C machines increased very rapidly, and in **1962** one manufacturer produced a point-to-point N/C drilling machine with locating accuracy of ± 0.001 in./ft at a price under $10 000.

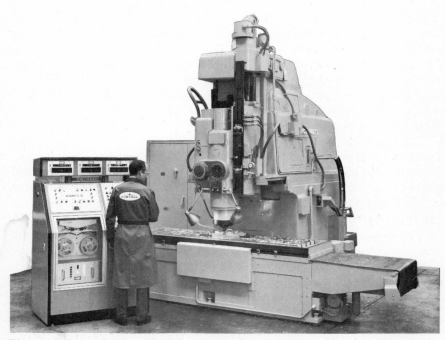

Fig. 1-1 One of the earliest N/C machines to use integrated circuitry. A heavy-duty production milling machine. (*Cincinnati Milacron.*)

* N/C Yesterday, *American Machinist,* vol. 108, no. 22, p. NC 6, 1964.

During these years, and continuing today, the electronics industry was busy. First miniature electronic tubes were developed, then solid-state circuitry, and then modular, or integrated, circuits (Fig. 1-1). Thus the reliability of the controls has been greatly increased, and they have become more compact and less expensive. At the same time intricate control systems are being developed to do jobs and perform functions which had never been thought possible.

N/C Equipment Today

The list of numerical control machines is constantly growing, with several hundred sizes and varieties of machines, many options, and many varieties of control systems available. The sizes range from small 18-in.-square tables to "skin mills," with tables up to 6100 mm [20 ft] wide and over 30 000 mm [100 ft] long (Figs. 1-2 to 1-4).

> *Note:* Some of the words used in the descriptions of the N/C machines illustrated in this chapter probably do not mean much to you now. However, as you study further, you will find that the vocabulary of numerical control

Fig. 1-2 A three-axis Numerimill N/C profile milling machine for close-tolerance milling of even the superalloys. Table travel X—3650 mm [144 in.], Y—up to 4000 mm [156 in.], Z—460 mm [18 in.]. Up to 30 kW [40 hp] on each spindle. (*Giddings & Lewis Machine Tool Company.*)

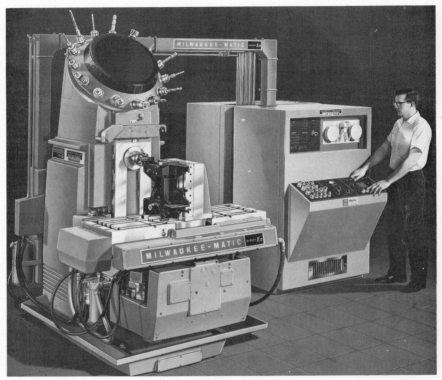

Fig. 1-3 A medium-priced three-axis horizontal-spindle milling machine, with a 15-tool turret and indexing table. Uses variable block, word address. CNC is available. (*Kearney & Trecker Corp.*)

equipment soon becomes familiar. Then these illustrations, and the others in the book, will furnish you with a valuable knowledge of the variety of N/C machines which are available.

Frequently, the mention of tape-controlled machines brings to mind the metal-cutting machines which will drill, tap, bore, and mill. However, there is a constantly widening range of equipment using tape control.

The N/C **engine lathe** and **turret lathe** are being made by several manufacturers (Figs. 1-5 and 1-6). Tape-controlled turret and quick-change **punch presses** are finding wide acceptance (Figs. 1-7 and 1-8).

Other numerical control machines are **tube benders, flame cutters, and riveting machines. Tape-controlled drafting machines** are now quite commonplace and are being used for engineering drawings, as well as for checking complex N/C tapes (Figs. 1-9 and 1-10).

Fig. 1-4 An older machine, but with Hydro-sense, which senses contact of a drill or tap with the workpiece and automatically changes to the programmed feed rate. Uses simple WORD ADDRESS programming, with floating zero. (*Giddings & Lewis Machine Tool Company.*)

Fig. 1-5 A heavy-duty N/C lathe, up to 37-kW [50-hp] motor. Switchable inch/metric programming. Feeds in either mm/min or mm/rev [ipm or ipr]. Used for chucking or between-center work. (*Lodge & Shipley Company.*)

Fig. 1-6 A modern slant-bed N/C turret lathe. Up to 458-mm [18-in.] chuck, and 45-kW [60-hp] motor available. TNC, program edit, etc., also available. (*Jones & Lamson Division of Waterbury Farrel.*)

Fig. 1-7 A small, single-punch N/C Fabricator, with 760-mm [30-in.] throat and 267-kN [30-ton] capacity. For punching, notching, etc. WORD ADDRESS programming. Now discontinued, but many still in use. (*Strippit Division, Houdaille Industries, Inc.*)

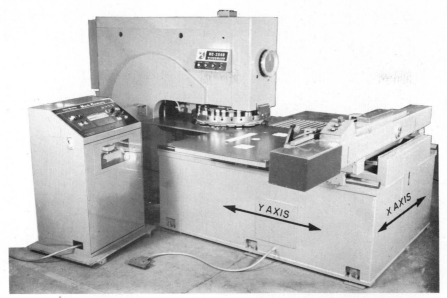

Fig. 1-8 An N/C turret hole-punching machine. Made from 66.75- to 445-kN [15- to 100-ton] capacity. Uses WORD ADDRESS, and up to 30 different punches called in any order. Handles 900 × 1800 mm [36 × 72 in.] sheets. (*Wiedemann Division, Warner & Swasey Co.*)

Fig. 1-9 This N/C drafting machine reads punched tape, IBM cards, or magnetic tape as options. The stylus will draw at 15.2 m/min [600 ipm] with accuracy of ±0.18 mm [0.007 in.]. (*Gerber Scientific Instrument Co.*)

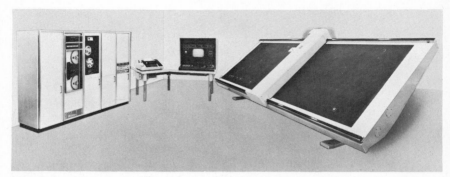

Fig. 1-10 An N/C drafting machine which is controlled by a computer. Table sizes up to 1.5 × 7.3 m [5 × 24 ft]. Scales data from 0.100 to almost 1000 times actual size. Up to 19 m/min [750 ipm] drafting speed. (*Gerber Scientific Instrument Co.*)

N/C **wire-wrap** machines make it possible to wire quickly and correctly many varied systems of computers and other electronic equipment and N/C machines **insert components** into circuit boards. These circuit boards may have been drilled by a tape-controlled machine with several drill heads, each with multiple spindles (Fig. 1-11).

Inspection machines can accurately check the inside and outside contours of curved shapes or center-to-center distances of holes. Other fields

Fig. 1-11 A tape-controlled N/C printed-circuit drilling machine. Drills four stacks of circuit boards with up to four boards in each stack. Uses WORD ADDRESS programming. Spindle speeds up to 35 000 rpm. (*Edlund Division, Monarch Machine Tool Co.*)

in which tape control is being used or explored include programmed inspection of electrical and electronic equipment, welding, lasers, electron-beam cutting and welding, grinding, and die sinking. New applications are constantly being found.

The Common Factors in N/C

With such a bewildering array of equipment available, it might seem to be hopeless to try to simplify the job of writing programs and making tapes to control all these different machines. Fortunately, many of the N/C machines for all kinds of work **can** be hand-programmed quite simply, and the efforts to standardize the tapes and control commands have been quite successful. The basic method of calling for machine action is similar for practically all tape-controlled machines. For example, the taped command to move 250 mm along the X axis is basically the same whether the N/C machine is a drill, welder, milling machine, or lathe.

The numbers and letters punched into the tape can tell the N/C machine how far to move, in which direction to move, and, in many machines, how fast to do it and which cutting tool to use. This seems quite complicated (and sometimes it is), but the combined efforts of the machine builders, the control manufacturers, and the standards groups have greatly simplified much of the work.

Reliability Today

Good machinists can work to a tolerance of ± 0.02 mm [0.001 in.] on an old and worn machine if they "know" the machine. Sometimes no one else in the shop can work so close on that particular machine.

With the arrival of numerical control, this skilled hand control is no longer possible; in fact, there are no hand wheels on most N/C machines. Thus the builders of N/C machines soon found that they had to make these new machines extremely rugged and extra rigid, with very low friction on the slides and with backlash practically zero.

In response to these requirements, new types of slides, such as recirculating ball bushings and machine tables riding on a pressurized film of oil, are being developed and used. Spindles are frequently larger, bearings are heavier, and lead screws are more accurate. The average N/C machine is cutting metal a high percentage of the time, and thus is subjected to accelerated wear and must be built to take it.

The **control systems** of N/C machines are designed to give signals to the machine in units of 0.02, 0.005, or 0.002 mm or in units of 0.001, or 0.0002, or 0.0001 in. The manufacturer guarantees accuracy of ± 0.12 to ± 0.002 mm [± 0.005 to ± 0.0001 in.] for metal-cutting operations. A few machines can be purchased with systems capable of control to 0.0002 mm [0.00001 in.]. The repeatability (that is, the comparison between the dimensions of each piece produced) is even closer, usually about half the positioning tolerance. Of course, the extremes of accuracy are obtained at a much higher cost, and not all companies need these "sophisticated" machines.

These figures of accuracy and repeatability assume that the machine is **properly cared for** and is in a location reasonably free from excessive vibration, dirt, and temperature changes. Even the best machines will not operate properly with a heat-treating furnace on one side and a punch press on the other. The conditions which affect the accuracy of any machine tool are shown in "What Affects Machine Accuracy," below. These become especially important when numerical control machines are used.

Proper maintenance of N/C equipment is of special importance. The higher cost, the higher percentage of metal-cutting time, and the automatic control of N/C machines make it necessary to be extra certain to follow good housekeeping, proper lubrication schedules, and careful preventive maintenance. These items, of course, apply to all equipment in a shop, and they are no more complicated for numerical control machines than for any high-precision equipment.

Purchasers of N/C machines may be concerned about the maintenance of electronic circuits, since they have seldom used them. Today, however, with simplified solid-state circuits, quickly replaced "cards" for maintenance personnel, and high-quality mechanical systems, it is not unusual for a shop to report as low as 5 percent down time on its tape-controlled equipment.

The machine manufacturers will train the user's repairer in one or two weeks so that he or she can quickly find and correct most difficulties. Troubleshooting manuals, special lights and contacts built in for diagnosing trouble, and special diagnostic service equipment are supplied, or can be purchased from the manufacturers. Thus even the less expensive numerical control machines are designed to be thoroughly reliable and very accurate pieces of equipment.

What Affects Machine Accuracy?*

The builder can see to it that built-in geometry, type of construction and similar factors are such that squareness, spindle concentricity, and surface flatness are maintained to the highest degree practicable. Design is also largely responsible for effective dissipation of heat generated internally by friction, motors and electronics—all of which can affect the accuracy of a machine's output. But from then on, holding tolerances is up to

* N/C—The Second Decade, *American Machinist,* vol. 108, no. 22, p. NC 12, 1964.

the user, and here are some of the factors that affect accuracy on the shop floor:

1. Flatness and solidity of the foundation.
2. Vibration from nearby equipment, or from interrupted cuts.
3. Workpiece (and fixture) weight and cutting loads that are such as to contribute to machine distortion.
4. Uniformity (or lack of uniformity) and rate of change in temperature of air surrounding the machine—affected by sunlight, cooling fans, ventilators, and the like.
5. Plant services (lubrication and maintenance, air pressure, plant voltage level and fluctuations).
6. Size, sharpness, material, deflection and geometry of cutting tools used on the machine.
7. Workpiece material and geometry (machinability, stiffness, and overhang).
8. Workholders.
9. Sequence of operations (part heating, reduction in rigidity, chip clearing).
10. Duty cycle (as it affects temperatures).
11. Measurement accuracy.
12. Age and wear of machine.

Chapter 2

WHAT IS N/C PART PROGRAMMING?

There are a few basic ideas which should be understood before studying the actual programming in detail. This chapter presents in broad outline the two types of control systems, the axis rotation, and the preliminary work needed before writing a numerical control part program manuscript. Later chapters include more detailed consideration of some of the items in this outline.

There are two principal types of control systems for N/C machines. The more common is the **point-to-point** positioning system, sometimes referred to as numerical positioning control (NPC). The other is numerical contouring control (NCC).

Point-to-Point Programming (NPC)

The function of the NPC positioning system is to move the machine table or spindle to a specified **position** so that machining operations may be performed at that point. **The path taken to arrive at this point is unimportant** (except to avoid collisions) in a point-to-point system, since no cutting is done until **after** the point has been reached. As this movement from one point to the next is non-machining time, it is made as rapidly as possible, usually at a rate of more than 2500 mm/min [100 ipm]. 200 I Pm

As shown in Fig. 2-1, several paths may be taken between the two points A and B (actually, path 2 is the most commonly used). If the job is to drill an 8-mm hole at point A and another 8-mm hole at point B, the machine operator does not need to be concerned with the path used to reach the points. Of course, the shortest path takes the shortest time, so paths 3

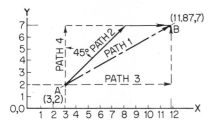

Fig. 2-1   Four possible straight-line paths from point *A* to point *B*.

and 4 would not be desirable; and most point-to-point machines cannot take path 1, as shown later.

Straight-Cut, or Picture-Frame, Milling (on NPC Machines)

In order to do simple milling operations it is only necessary to add equipment to control the rate of feed from one point to the next. This milling feed control is either standard equipment or readily available on most of even the lower-priced numerical control machines, though a few excellent machines have either limited or no milling capabilities. However, with most point-to-point systems, the only direction which is **accurately** controlled is in straight lines—left, right, forward, and back, as shown in Fig. 2-2. If these directions are made in sequence, they will form a square or rectangle, making a cutter path resembling a "picture frame," as in Fig. 2-3. Some N/C programmers do occasionally use the 45° path for milling, with a modified feed rate, where accuracy is not essential.

Thus a great deal of useful milling can be done on a point-to-point machine. An example of some of the cuts that can be made is shown in Fig. 2-4. Notice that cutting around the outside corners of the part will make a sharp corner, but the inside corners will have the same radius as the cutter. Chapter 7 describes in detail how milling is programmed.

Continuous-Path, or Contouring, Program (NCC)

A contouring (NCC) control system must have independent control of the speed of each of the driving motors, and thus be able to regulate the rate of travel of the table (or spindle) on at least two axes at the same time.

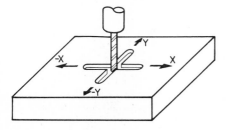

Fig. 2-2   On a simple point-to-point N/C machine (without linear or circular interpolation) a milling cutter can be controlled only on straight lines along the *X* and *Y* axes.

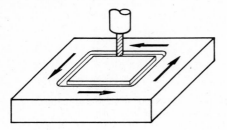

Fig. 2-3   An example of picture-frame, or straight-cut, milling.

Assume that a slot is to be cut from left to right at 30°, as shown in Fig. 2-5. Notice that as the table (or spindle) moves 86.6 mm to the right (X axis), it must move 50 mm in back (Y axis), and will travel 100 mm on the diagonal (hypotenuse). Do you remember that 0.866 is the cosine of 30° and 0.500 is the sine of 30°? This type of cut requires the two driving motors (X and Y) to run at unequal speeds, and this means that quite an elaborate control-and-drive system is needed. This contouring control system is, of course, more expensive than the point-to-point equipment. Control of the rate of travel in two directions, proportional to the distance moved, is called **linear interpolation.**

This same control system can cut curves to very close tolerances by cutting a series of straight lines—either chords, tangents, or secants—as shown in Fig. 2-6. It is not too difficult mathematically to compute the X and Y coordinates of the ends of each straight line around a circular path

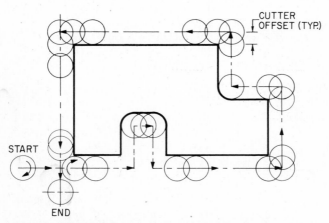

Fig. 2-4   Some milling cuts which can be done on most point-to-point (NPC) N/C machines. Conventional (up) milling shown with travel in CCW direction. Notice that going around the part in clockwise direction will result in climb (down) milling.

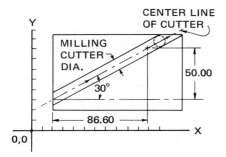

Fig. 2-5 One type of contouring cut which requires linear interpolation.

(Appendix A), but it is quite time-consuming. For example, to cut chords or tangents around a 100-mm-dia. circle with 0.13 mm maximum offset requires 45 points, which make chords about 7.19 mm long. If only 0.02 mm offset is allowed, it requires 100 points, with chords 3.20 mm long. Thus this work is frequently done on a computer, using one of the N/C program languages, such as APT or AD-APT, which will give coordinates and other machine commands. Other computer programs may be used just to get the X and Y coordinate values, which will then be used to write the program.

Continuous-path contouring will be described in Chap. 13, on Computer Programming of N/C Tapes.

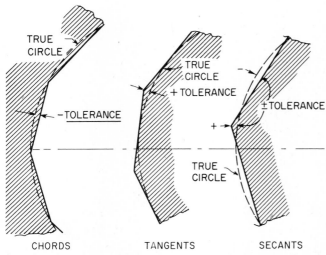

CHORDS   TANGENTS   SECANTS

Fig. 2-6 Three possible straight-line approximations of a circle. Lines may also be fitted to noncircular curves. Actual lengths of cut are approximately 0.38 to 6.4 mm [0.015 to 0.250 in.] long on the average part.

The N/C Machine Axes

Tape-controlled machines are spoken of as two-, three-, four-, or five-axis machines. How can this be, when we know that this is a three-dimensional world? The answer is that in N/C the word "axis" refers to any direction of motion which is fully controlled by tape commands.

By looking into the corner of a room, the three familiar axes may be seen. The side going to your left or right is the X axis. The side going toward and away from you is the Y axis. The up and down is the Z axis. In a **two-axis** N/C machine, only the X- and Y-axis motions are controlled by **tape commands.** All other machine movements (such as raising and lowering the spindle and setting depths) are done by hand or by mechanical, air, or hydraulic controls.

When the depth (Z axis) of cut is controlled by the dimensions given by the tape, the machine is a true **three-axis** N/C machine. As will be shown later, there are N/C machines in which the tape calls for a choice of preset depths. This is **not** true three-axis control, though it does give tape-actuated Z motions (Fig. 2-7).

The **fourth axis** may be the machine table to which the workpiece is secured. This table may rotate at a controlled rate **while the work is being machined.** Some tape-controlled machines have **indexing** work tables with from 4 to 360 000 tape-controlled positions, but they do not always index while the work is being machined. These indexing tables are of great value, but do not make a true four-axis machine.

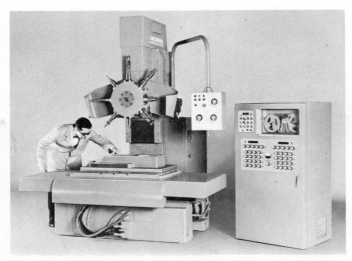

Fig. 2-7 An eight-spindle N/C drilling-milling machine. With tape selection of spindles (T function), Z rapid (R functions) and tape-controlled speeds and feeds (F and S functions). Uses tool-length compensation, WORD ADDRESS, fixed zero, with FULL ZERO SHIFT available. (*Cincinnati Milacron.*)

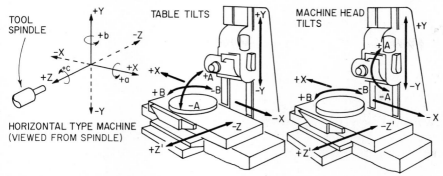

Fig. 2-8   Alphabetical notation of five-axis horizontal-spindle N/C machines. Z' motion is plus when work moves away from the spindle.

The **fifth axis** may be the ability to swivel the entire machine head at a controlled rate. This, of course, tilts the axis of the cutter so that it can be perpendicular to a tangent point on a curved surface. Two possible combinations of five axes are shown in Fig. 2-8 for horizontal-spindle N/C machines, and Fig. 2-9 for vertical-spindle N/C machines.

The construction and use of the fourth and fifth axes vary between different manufacturers. Some N/C machines are equipped with a **tilting table.** If this is simply an adjustable tilt, it is not a true **fifth axis.**

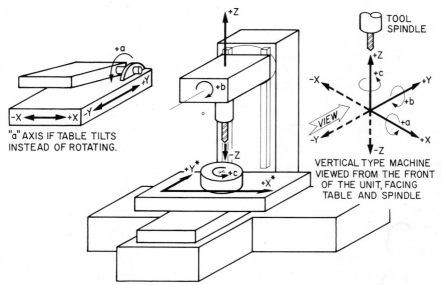

*See Fig. 4-2 and 4-3 explaining table and carriage + and – movements.

Fig. 2-9   Alphabetical notation on five-axis vertical-spindle N/C machines, showing spindle and spindle housing moving in Z axis. On some machines, the entire spindle column moves in the X axis.

Fig. 2-10   A five-axis N/C machine with automatic tool changing of up to 60 tools, table movement is X, carriage movement is Y, and spindle axis is Z. Also, 360° table rotation (C axis) plus 150° head rotation (A axis). Dual spindles, with 18.6-kW [25-hp] motor. Usually programmed by computer, using SPLIT or APT language. (*Sundstrand Machine Tool Division, Sundstrand Corp.*)

Notice in Fig. 2-8 that there are **six** possible motions. Rotation around the X axis is the a direction; rotation around the Y axis is the b direction; and c rotation is around the Z axis. This lettering system applies to both vertical- and horizontal-spindle machines. If observation is made looking in the "plus" direction from the (0, 0, 0) origin, all "plus" rotation is clockwise. Figure 2-10 shows a large five-axis N/C machine.

Programming for these four-and five-axis machines is not included in this book. It is not necessary to spend time learning the notation of the six axes at this point in the study of numerical control programming. As you progress, however, you may see these notations used, so it is valuable to be able, at least, to recognize their meaning.

Before going on to the specific details of manual programming, attention should be given to some of the things which must be considered when planning to machine a workpiece on a tape-controlled machine.

Steps in Planning for Programming of N/C (Fig. 2-11)

Ideally, all the steps shown below are followed in making a part on a tape-controlled machine. However, in smaller shops many of these steps are combined, and the N/C programming department may have to follow through on most of the items.

1. **The blueprint** of the part to be made. If the engineering and

drafting departments plan ahead for N/C, the dimensioning can be done in a way that simplifies programming and checking.

2. **Specifications** of material and desired finish. These are usually on the blueprint. They are quite important, especially in deciding feeds, speeds, and needed finish cuts.

3. **Lot size** and the **annual use** expected. These two items will influence the amount of time and money to be spent on fixturing and planning. It is worthwhile checking to see if it will be more economical to tool up to do the job on conventional machines, considering the complexity of the part and the amount of tooling needed.

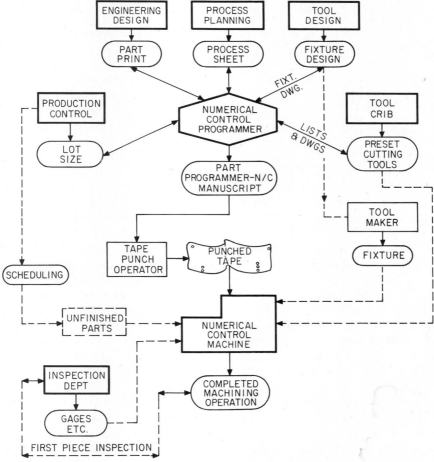

Fig. 2-11   Steps in planning an N/C program (for hand programming). Notice that many lines of communication go both ways.

An economical lot size for N/C may be from 1 to over 1000 pieces.

✳ 4. **Operation sheets,** or **process sheets,** are made showing all work that must be done to complete the part. These may include several sketches. This listing will include all N/C machining, conventional machining, inspection, burring, etc., as needed.

✳ 5. **Tooling** is listed, including all jigs and fixtures and, especially in N/C, all cutting tools. For numerical control machines, drawings are often made of the cutting tools. These drawings specify tool size, length, etc., to simplify tool changing and setup.

✳ 6. **The program manuscripts** for the N/C work can now be written by the programmer, using the information, drawings, and sketches made during previous planning.

✳ 7. **Punch the tape,** from the programmer's manuscript. This can be done by any typist after a short period of training. Chapter 3 describes the tapes and tape-punching machines. Computers can also punch a tape.

✳ 8. **Try out the tape.** One way that this can be done is by "jogging" through the tape on the machine with the work in place. If the workpiece is damaged, it may be patched with aluminum epoxy and used again. Some companies use a simulator, which is hand-operated, or a wooden or plastic workpiece can be used on the machine. The purpose of this tryout is to catch wasted motion and the larger errors which might cause loss of time or collisions.

✳ 9. **First-piece inspection.** When all large errors have been corrected, the first piece is run. At this time the operator (and often the programmer) keeps a close watch to see that feeds and speeds are correct. On practically all tape-controlled machines, the feed and speed can be changed by the operator during the run, with controls on the operating console. All necessary changes are later made on the tape.

The first piece successfully machined is taken immediately to the inspector for a thorough check. If the part is acceptable, the job is run. During production it is only necessary to check critical dimensions on each piece and occasionally check a complete piece, because the tape and the machine will control all locations very reliably.

All the foregoing steps needed to produce a good part on a tape-controlled machine lead up to, and result from, the work in step 6, the program manuscript. This manuscript is written by the N/C parts programmer. An N/C parts programmer, as shown in Fig. 2-11, should be able to understand the work of several groups. He or she then will combine their contributions with a familiarity with good machining practices and a specialized knowledge of numerical control programming method and will write the program manuscript.

The information in this book can supply important steps in learning this specialized knowledge of numerical control programming, which a numerical control programmer must have and which is valuable for all who work with N/C machines. This is not a difficult subject, and it is an important, interesting part of the job of successfully using the rapidly increasing number of numerical control machines.

Chapter 3

THE N/C TAPE
AND ITS CODING

Tape Materials

Most numerical control machines receive their signals from holes punched into a 1-in.-wide tape. This tape can be made of paper, paper-Mylar* "sandwich," aluminum-Mylar* laminates, or other materials. Most of these tape materials can be bought in a variety of colors, such as pink, yellow, green, and blue.

Paper tape is inexpensive. It sometimes has an oil- and water-resistant treatment which makes it quite durable, and some companies use it, especially for short runs. However, it is still fairly easily damaged, and optical tape readers (which use light and photocells) are sometimes sensitive to stains on the paper caused by oil, grease, or cutting fluids. Black paper tape is sometimes used for photoelectric tape readers. The aluminum-Mylar tape is more expensive, but is almost indestructible and is not affected by oil, etc. Therefore many companies have standardized on this as the material for the punched tapes used on the N/C machines in production.

All these tapes are purchased in rolls, frequently about 200 mm [8 in.] in diameter. These rolls contain from 300 to 600 m [1000 to 2000 ft] of tape, depending on the thickness of the tape.

Tape Specifications

No matter what the color or material, all N/C tape is 25.4 ± 0.076 mm [1.000 ± 0.003 in.] wide, and the usual thickness is 0.10 ± 0.008 mm

* Mylar is Du Pont Company's trade name for a very tough plastic.

[0.004 ± 0.0003 in.]. This provides for eight tracks (or channels) of punched coding holes plus the row of smaller holes which fit the tape-feeding sprockets.

The complete tape dimensions are shown in Fig. 3-1. This is the Electronic Industries Association (EIA) Standards RS-227. The same dimensions have been adopted by the Aerospace Industries Association of America (AIA) in their National Aerospace Standards (NAS-943 and NAS-955) for numerical control machines.

Magnetic Tape

It is possible to control an N/C machine with a magnetic tape. This is a plastic tape coated with an iron oxide, much the same as that used in tape recorders. The coding on these magnetic tapes is done on a computer, and instead of punched holes, small areas of the iron oxide coating are magnetized. When put onto a special tape reader, these magnetized areas cause signals to be sent to the N/C machine control system.

Magnetic tape can store many more signals per inch than the punched tape. Thus it is much shorter. It can also be run through the tape reader much faster. However, magnetic tape requires careful handling since it is easily damaged by any electric or magnetic fields. A magnetized screwdriver placed too close could ruin the tape's usefulness. Dirt or oil on this tape can also cause it to make an error in reading.

Thus, in the far-from-antiseptic surroundings of most machine shops, the EIA coded punched tape may be more practical. Magnetic tape is, of course, widely used in the computer field and on some N/C equipment.

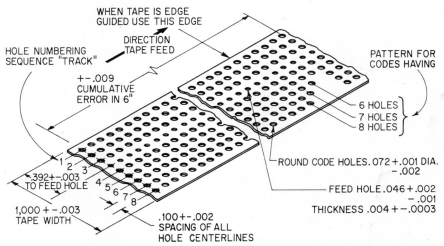

Fig. 3-1 Standard 25.4-mm [1-in.] perforated tape, according to EIA and NAS specifications.

Punching the Tape

In manual, or hand, programming the numerical control programmer writes a program manuscript, which is a list of all the signals the N/C machine will need in order for it to do the work required. The next few chapters teach you how to do this.

After the manuscript, or N/C program, is written, the codes for each letter and number must be punched into a tape. It certainly would not be efficient if someone had to punch each individual hole in the proper place. Several machines are available to speed up the job.

Tape-punching equipment is made by several companies. The Flexowriter (Fig. 3-2), made by Friden Division of Singer, is no longer manufactured, but there are hundreds still in use. A more modern tape punch, made by Artec International, is shown in Fig. 3-3. Other companies also make advanced versions of tape-punching equipment (Fig. 3-4). The elec-

Fig. 3-2 Flexowriter automatic writing machine, used for punching and duplicating N/C tapes. No longer made, but hundreds still in use. (*Friden Division of Singer.*)

Fig. 3-3 A modern typewriter of the "Selectric" type, equipped to punch, read, and duplicate tapes up to 1080 characters per minute. Uses solid-state components. (*Artec International Corporation.*)

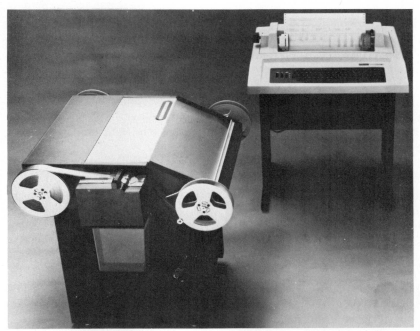

Fig. 3-4 A microprocessor-based N/C programming terminal. A buffered keyboard allows 100 characters to be typed and corrected before the tape is punched. Provides search, and error messages, and inch/metric conversion. (*Numeridex Incorporated.*)

tronic circuits of these machines enable them to punch the proper code in the tape for every key on the tape-typewriter keyboards. These punched codes include plus and minus signs, carriage return, tabulate, space, upper and lower case, and all the usual symbols, such as parentheses and asterisks.

Usually, while the tape is being punched, a typewritten copy is also being made. The same machines can duplicate a tape, so that one copy may be filed and the other sent to the shop, and can also type out the words or numbers from a tape at well over 100 words per minute. The Flexowriter can "verify" a tape by comparing an existing tape to the characters being punched while typing a duplicate tape.

Although it may seem wasteful to type a program twice, the discovery of a typist's error before the tape is placed on the N/C machine may avoid spoiling a valuable part or breaking a tool. Of course, if the tape is made from a computer program, it is not checked (verified) because the computer almost never makes a typographical error. Another way to check a tape is to double-space the first typeout and retype on the space exactly below. By comparing the two typeouts, errors can be seen very clearly.

This combination of capabilities means that these tape-punching machines can be used for punching tapes for almost all varieties of coding systems. (The only exception is the binary code, discussed later in this chapter.) These same "typewriters" are used for office work of many kinds.

A separate machine for comparing and duplicating tapes is shown in Fig. 3-5. This machine will quickly and accurately verify tapes, and even make a good continuous tape from a spliced one, and omit the delete codes used for corrections.

Correcting the Punched Tape

If, on any tape-punching machine, a typing error is made, it can be erased by using a DELETE key which punches holes in the first seven tracks of the tape. The correction is then typed, following the delete codes. When a duplicate tape is made, the delete is not copied. Numerical control machines will skip over all the delete codes and use the corrected information.

Engineering design changes in order to find a better method, or desirable changes in feeds and speeds may require that added information be put on the tape or that a long section of tape be changed. To do this, the new or corrected information is punched out on a separate section with about 75 mm [3 in.] of code delete at each end, and this is spliced into the original tape. The procedure is similar to that used to splice motion-picture film. The splicing tape is usually very thin, since some tape

Fig. 3-5 An off-line unit which will verify and duplicate N/C tapes at 120 characters per second. Switches allow the operator to correct errors, delete, or add new characters manually. (*Tally Corp.*)

readers will not handle much over the 0.10-mm [0.004-in.] thickness. If there are many splices, it is usually wise to make a completely new, unspliced duplicate tape as soon as possible since the splices may not be as strong as the tape and thus may break if used many times.

TAPE CODES

There are several systems of arranging the holes in a tape to code the numbers, letters, and symbols on a typewriter keyboard. These systems use from five to eight tracks (sometimes called channels, or levels) of holes. In numerical control work, eight-track tape and two coding systems are used.

The Binary Code and Binary-coded-Decimal Code

The most positive way to control an electrical circuit is to have it either *on* or *off*. If signals are to be sent to this type of circuit, only two signals may be used, and a way of specifying these signals is needed. It is usually agreed that, to specify these signals in writing, a zero means "keep this circuit off" and a figure 1 means "turn this circuit on."

In a tape with punched holes, the zero is indicated by no hole and the one is indicated when a hole is punched in the tape. With this limited amount of information, numerical control equipment must be able to send signals involving letters and numbers. Since only two signals are available, the system is called the **binary** system ("bi-" meaning "two").

The original binary system starts at the right with zero or one, and as it moves to the left, it raises the digit 2 to a higher power. In algebra you learn that

$$2^0 = 1 \qquad \text{zero power}$$
$$2^1 = 2 \qquad \text{first power}$$
$$2^2 = 2 \times 2 = 4$$
$$2^3 = 2 \times 2 \times 2 = 8$$
$$2^4 = 2 \times 2 \times 2 \times 2 = 16$$

This binary coding was once used. However, it is a difficult code to read and to write. For example, the number 2673 would be written as shown in Fig. 3-6.

Position or track	12	11	10	9	8	7	6	5	4	3	2	1
Power of 2	2^{11}	2^{10}	2^9	2^8	2^7	2^6	2^5	2^4	2^3	2^2	2^1	2^0
Quantity represented	2048	1024	512	256	128	64	32	16	8	4	2	1
Example: 2673 =	1 2048	0 +0	1 +512	0 +0	0 +0	1 +64	1 +32	1 +16	0 +0	0 +0	0 +0	1 +1

Example: 2048
                512
                 64
                 32
                 16
                  1
Total   2673   In binary code this would be written 101001110001.

Fig. 3-6 The binary system of coding numbers.

Binary-coded Decimal

A simpler method, called **binary-coded decimal,** usually abbreviated BCD, is most frequently used. This system uses only the first four positions of the binary code, quantities 1, 2, 4, 8 (Fig. 3-6). Notice that the four numbers 1, 2, 4, and 8, added together as needed, will make all the numbers from 1 to 15. For example, 8 + 4 + 2 + 1 = 15.

> *Note:* By adding the fifth position (2^4 = 16), numbers up to 31 may be coded. This fifth-position BCD is used in coding tool numbers for some N/C machines.

The "decimal" part of the system means that each digit (a digit is a single number from 0 to 9) will be written by the programmer, and read by the N/C machine, in order from left to right. This is the way quantities are usually written. For example, the number 92 673 is written in BCD as shown in Fig. 3-7.

Our decimal system uses only the digits from 0 to 9; so even though the four positions 1, 2, 4, 8 can add up to 15, the binary-coded-decimal system does not use the numbers from 10 to 15. The tape reader reads one digit at a time and, with special electronic circuitry, places each number in its proper decimal position. Figure 3-8 shows the BCD codes approved by the EIA and AIA plus other codes which are occasionally used.

ASCII Code

Another code which is now available on some N/C machines is called the U.S.A. Standard Code for Information Interchange, often abbreviated ASCII (pronounced "asky"). This code was compiled by a large committee from many groups working with the United States of America Standards Institute (now named American National Standards Institute).

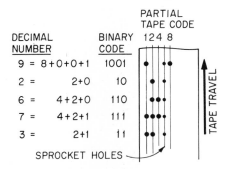

NOTE: This is only <u>part</u> of the actual
  N/C Code, used to illustrate the basis of
  Binary Coded Decimal coding.

Fig. 3-7 Some binary-coded-decimal (BCD) numbers punched into a tape. No parity check shown.

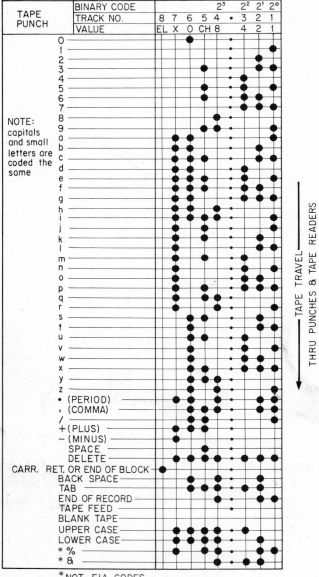

Fig. 3-8 The EIA RS-244 "character codes for Numerical Machine tool control perforated tape." (*Electronic Industries Association.*)

The objective is to have one coding system which will be, as nearly as possible, an international standard for all information-processing systems, communications systems, etc. The basic code is shown in ANSI Standard X3.4-1968 (the "1968" shows the latest revision, since it may be changed from time to time).

The 10 digits in this ASCII code are the same as in the BCD coding, but with holes punched in two additional tracks, to identify the numbers and certain symbols.

The ASCII code X3.4 provides coding for both capital and small letters, which is, of course, not true in the present BCD tape codes. The ASCII letter codes are quite different from those used in EIA RS-244.

Many N/C control units include the necessary electronic equipment to handle both the BCD coding and the ASCII coding. A simple switch provides a choice of coding, so that the machine may be run with tapes punched in either code.

Parity Bit

A bit is a mark, 1 or 0 in the binary numbering system. In punched tape it is a hole or the absence of a hole. In the chart of Fig. 3-8, notice that column 5 is labeled "CH," meaning **check** (or parity check), and that this column has a hole punched in it, in addition to the holes needed for the BCD code for some of the numbers. This is a very necessary addition.

The equipment for punching and reading tapes is extremely reliable. Nevertheless, it is possible for errors to be made. There is no simple system for checking that the numbers read are 100 percent correct. However, the EIA specifies a very efficient method of eliminating most errors which might be caused by blocked-up or unpunched holes, which would cause an error in reading one row. The method of checking is simple: **there must be an odd number of holes across every row in the tape.**

Notice on the chart in Fig. 3-8 that the numbers 3, 5, 6, and 9 need only **two** holes punched in the tape to add up correctly—and two is an **even** number. Therefore column 5 is used as a check (CH, or a parity bit), and this hole, or bit, is punched, in addition to the regular numerical codes, in order to make an odd number of holes (1, 3, 5, 7) in every row of the tape. This is referred to as an odd-parity system. "Parity" means "equality, as in amount, status, or character." Most N/C tape control equipment has electronic controls which will check for this odd parity and stop the machine if an even number of holes is read into the system.

Note: The ASCII coding uses an even-parity check.

Coding for Zeros

In basic BCD, zero would be coded by not punching any holes in the row. This would be awkward, since there are times when it is necessary to just run the blank tape through the reader, and no signal to the MCU is desired. Thus, in the Electronic Industries Association (EIA) RS-244 coding system for punching tape, the standard code for zero is a single hole punched in track 6, as shown in Fig. 3-8. We have added two codes to the official list, because these are used in some N/C machines. These are the percent (%) and ampersand (&) symbols. Their use is described in Chap. 5.

When you know the meaning of the holes in the first six rows of the tape, it is quite easy to learn to read the numbers on a punched tape.

There are five codes that are used quite frequently which should be learned. Refer to the chart of Fig. 3-8 while reading about them.

Plus and Minus Signs

Two frequently used symbols are the plus (+) sign (tracks 5, 6, and 7) and the minus (−) sign (track 7). Notice that a minus sign can easily be changed to a plus sign if an error has been made, but not the reverse.

Tab Code

The tab code is required by some N/C machines to identify and arrange the information in proper order. This code is punched in tracks 2, 3, 4, 5, and 6, and it is easily recognized when reading a tape. "Tab" is an abbreviation of "tabulate" (to arrange in columns), as in a "table" of figures. On an office typewriter this key jumps the carriage to a preset position so that all figures will be in line, one above the other. On a tape typewriter the same thing happens, for the same reason; and at the same time the tape is punched with the five-hole tab code.

End of Block

Track 8 has only one use, and in BCD it is never combined with holes in other tracks: this is, on the special electric typewriters, the carriage return. Its meaning in numerical control is "end of the line" (EL) or "end of block" (EOB); that is, the reading of this hole in track 8 tells the machine that it has received all the coded information necessary to perform the next job, and that the machine should follow these instructions immediately.

Rewind Stop Code (End of Record, EOR)

This code (tracks 4, 2, 1) is sometimes placed at the beginning of the N/C information on a numerical control tape. After a length of tape has been run through the reader and the machining has been completed, a signal is often given which causes the tape to be rapidly rewound, ready to be used for the next part.

This "stop code" stops the rewinding just ahead of the first tape command so that when the machine and tape are started for machining the next piece, the complete first block is read. Other ways of stopping the rewind are also used, so check the programming manual for your machine.

Leading and Trailing Zeros

Numerical control machines use five- to eight-digit "words" for specifying dimensions. For example, a 75.5-mm dimension may be written as shown, or as 75.500 or 075.50, and a 0.42-mm dimension may be written as 00.420 or 000.42, depending on the capabilities of the N/C machine. However, a **decimal point is never punched into an N/C tape.** Thus the above numbers would show on the actual tape as 75500, 07550, 00420, etc.

The machine control unit is frequently designed so that some of the zeros in a number may be omitted. Zeros before a number, as in **000.05**, are called **leading zeros.** Zeros after a number, as in 75.**500,** are called **trailing zeros.** Thus there are three possible ways to punch numbers into the N/C tape, as shown in Table 3-1.

1. **A number in every position.** This system can always be used, as it will be accepted by any N/C machine. In the second column of Table 3-1, notice that all five spaces are filled, using zeros to fill spaces before or after the significant digits. The same will apply for six-, seven-, or eight-digit machines. In a few tape-control systems all positions **must** be filled.
2. **Omit leading zeros.** N/C machines which use this system are designed so that the electronic controls assume that the last number

Table 3-1   EXAMPLES OF CODES OMITTING LEADING AND TRAILING ZEROS*

Dimension wanted on tape, mm	All places filled	Leading zeros omitted	Trailing zeros omitted
570.00	57000	57000	57
75.00	07500	7500	075
8.00	00800	800	008
0.25	00025	25	00025
0.05	00005	5	00005

* A similar system is used for inch dimensions, but with the decimal point after the second digit. Some N/C machines use additional digits—for example, 7426.00 for larger machines or 36.014 for machines working to very close tolerances.

punched for a command dimension is always in the hundredths or thousandths of a millimetre column, depending on the resolution of the machine control unit.

Thus, as shown in Table 3-1, the dimension 0.05 mm is programmed as the single number 5 when leading zeros are omitted.

3. **Omit trailing zeros.** This simplifies the punching of larger numbers like the 570-mm dimension shown in Table 3-1. Here the controls are arranged to accept the first number as always being the "hundreds" digit (or "thousands" digit in a larger machine).

THE STATEMENT BELOW IS IN WORD ADDRESS, VARIABLE BLOCK FORMAT.

     n 025   g81   X 97.00   Y 120.00   Z 5.00   m06   EOB

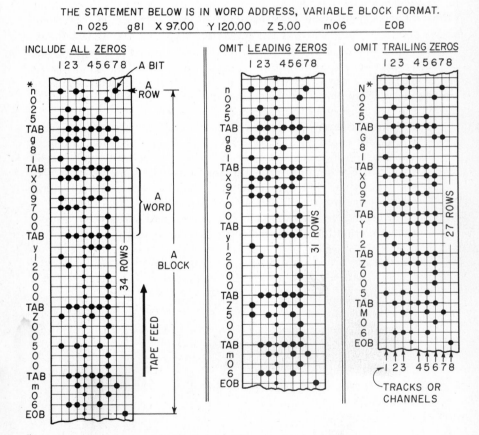

*The letters may be written as either capitals or small letters.
The punched coding is the same in BCD.

Fig. 3-9 Example of a full block of commands as they would appear on tapes punched according to each of the three methods of coding zeros, as shown in Table 3-1.

Both systems of omitting zeros are used quite widely and are easily understood with a little practice. Their use does shorten the length of tape needed for a program, and every "punch" omitted is one less chance of error. **Do not omit any zeros in the m and g codes,** which are to be studied in Chap. 6.

In writing a program, if in doubt, fill in all positions as shown in the second column of Table 3-1, and the code will be accepted. However, you **must** know whether the machine uses a five- or six-digit coding, or the punched numbers may be interpreted quite differently from what was intended.

A sample section of tape, punched in the three ways just described, is shown in Fig. 3-9. The meaning of the coding is explained in later chapters.

The line to be coded is

N025 g81 X97.00 Y120.00 Z5.00 M06 EOB

The tab is used between each word for convenience in listing, but it is not required in this code. The decimal points are, of course, not punched into the tape. The definitions of several important N/C terms are also illustrated in the figure. The definitions are as follows:

Bit is a single character. In a binary-coded-decimal system for punched N/C tape it is a punched hole or the absence of a punched hole.

Row is a line of holes perpendicular to the edge of the tape. In numerical control, each row represents a letter, a number, or a signal (such as the tab signal) to the machine control unit. Each row may contain one character. (**Character** is a letter, number, or symbol.)

Word is a set of letters and numbers (characters) which together give the N/C machine a **single** instruction. The word may be coded in a single row, such as EOB or EOR, or it may require several rows, such as X09700, shown in the example.

Block is a group of words which, as a unit, give the N/C machine the **complete** information for some kind of action. The block may have only one or two words or several words. It is all the information given between two END OF BLOCK characters.

Track, or channel, is the line of holes, or spaces, parallel to the edge of the tape, as shown in Figs. 3-8 and 3-9.

These definitions are included in the Glossary in Appendix G, but they should be learned now, since they are used often in the following chapters.

Chapter 4

THE BASIS FOR N/C DIMENSIONING

Numerical control programming uses the same system of cartesian coordinates that is used in plotting graphs in algebra. This system of coordinates is shown in Fig. 4-1. In algebra the X axis is usually shown as horizontal (left to right on the paper), and the Y axis is usually shown as vertical (up and down on the paper). With this view the Z axis would be illustrated by holding a pencil perpendicular to the paper and placing its point where the X and Y lines cross each other.

Where the X and Y axes cross is called the **origin,** or **zero point.** This same system of coordinates is used in numerical control programming. However, the X and Y axes are not always flat on the paper (or the machine table). This was shown in the diagrams in Chap. 2.

Quadrant Notation

A quadrant as used in the cartesian coordinate system is a quarter of a circle. For convenience the quadrants are numbered, and the numbering is counterclockwise (CCW), as shown in Fig. 4-1. This same quadrant notation is used in drafting and mathematics.

Notice that the plus and minus signs in this system indicate a **direction** from the zero point, along the X or Y axis. If the Z axis were included, plus (+) would mean a direction **above** the paper (the paper is the XY plane), and minus (−) would mean a direction **below** the XY plane.

The plus and minus signs, in some numerical control programming, are also used to indicate the direction of **travel** between two points. This use of signs is discussed later in greater detail.

In numerical control, it would be easiest if all work were done in the

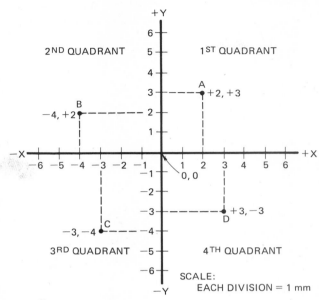

Fig. 4-1 Cartesian coordinate system.

first quadrant, since then all values of X and Y would be positive and no signs would be needed. Actually, much N/C work is in the first quadrant. However, any of the four quadrants may be used on different machines. Thus a programmer must be familiar with the use of plus and minus signs in all four quadrants.

Point Location—Two-Axis Tape Control

Some numerical control machines use tape commands for locating points in only the X and Y directions. These are called **two-axis** machines. On these machines, distances in the Z direction are controlled by the operator or by preset stops, as is often done on a drill press. Some two-axis N/C machines control the Z depth by the use of a system of cams, or several hand-set micrometer stops, which can then be selected by tape command.

In Fig. 4-1 points A, B, C, and D are shown, and their locations are shown in algebraic notation. It is universally agreed that in this kind of notation the X dimension is written first, then the Y dimension, and then the Z dimension.

Thus $A = +2, +3$ (usually written as $A = 2, 3$) means

Point A is located at $X = +2$ mm, $Y = +3$ mm, from the zero point.

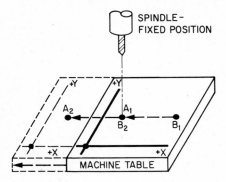

Fig. 4-2 The machine table moves **left** to position the cutter from point A to point B. This is the most frequently used method, especially in two- and three-axis N/C machines.

Point C is in the third quadrant, to the left of the zero point. This is indicated by the minus sign before the X dimension (-3). Point C is also below the zero point, and this is shown by the minus sign before the Y dimension (-4).

Thus $C = -3, -4$ means

Point C is located at $X = -3$ mm, $Y = -4$ mm, from the zero point.

Notice that either the machine **table** can move, as on the machine in Fig. 4-2, or the machine's **spindle** can move, as on the machine in Fig. 4-3. As shown in Figs. 4-2 and 4-3, this changes the actual direction of the motion, but the movement of the cutter as related to the work remains the same. This is taken care of by the makers of the machine and its controls. The programmer only has to specify the correct dimensions and the proper plus or minus signs for the hole or point location in relation to the established zero point.

In numerical control programming the plus signs may be written beside all positive locations. However, to simplify the work, it has been

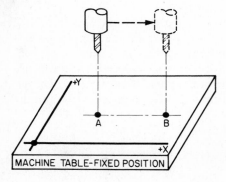

Fig. 4-3 The spindle moves **right** to position the cutter from point A to point B.

agreed that if no plus or minus sign is shown, the number is positive (+), though some N/C machines do require that the plus sign be used. Once again, this is the same notation used in algebra and other mathematics.

N/C Use of This Notation

This same method and notation are used for locating points in numerical control, except that the equals sign (=) and the decimal point are left out, and zeros are added to show where the decimal point is. In the following examples the decimal points are shown for clarity; they would not be punched into the N/C tape.

One way of listing the locations of the four points shown in Fig. 4-1 for use by a tape-controlled two-axis N/C machine is given in Table 4-1.

The exact method of arranging the figures for use for tape-controlled machines is shown in later chapters.

Z-Axis Notation

The addition of the Z axis is somewhat complicated by the fact that N/C machines are made with the spindle axis **vertical** (as in a drill press) and also with the spindle axis **horizontal** (as in a horizontal boring mill); so it has been agreed by the EIA and the AIA and the machinery builders that

A line through the center of the machine spindle is the Z axis.

The plane formed by the X and Y axes is, as in algebra, perpendicular to the Z axis.

To visualize this, draw the X and Y axes on a piece of paper. The paper now represents the XY plane, or surface. Use your pencil as the Z axis, and hold the pencil point against the paper at a 90° angle and on the intersection of X and Y. The result is as shown in Fig. 4-4a. Figure 4-4b shows from a slightly different angle how this notation would apply to a part located on an N/C machine using first-quadrant dimensioning.

Now hold the paper up so that it is edgewise to the table and the X axis is horizontal. The XY plane is now at 90° to the table. Hold the pencil,

Table 4-1 COORDINATE DIMENSIONS OF THE POINTS IN FIG. 4-1

Point	X location	Y location
A	2.000	3.000
B	−4.000	2.000
C	−3.000	−4.000
D	3.000	−3.000

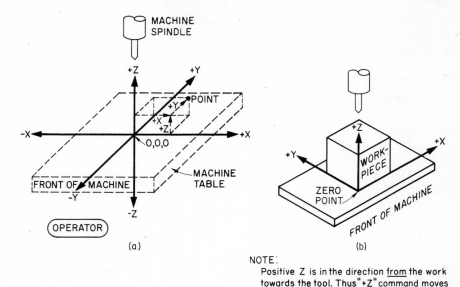

Fig. 4-4 Two ways of showing the standard *X*-, *Y*-, and *Z*-axis notation for verticle-spindle N/C machines.

representing the machine spindle, horizontally (at right angles to the paper), and touch the pencil point to the intersection of the *X* and *Y* axes. The pencil now represents the *Z* axis. The three axes will then be as shown in Fig. 4-5*a*.

Figure 4-5*b* shows how this notation is used on a part located in the first quadrant. Notice that this view is as if you were looking **outward from the spindle** toward the work, and from this view + *X* is to your right. This is according to the AIA Standard NAS-938 and EIA Standard RS-267.

Four- and Five-Axis Notation

In four- and five-axis N/C machines, this procedure becomes more complex, so other axes are given letters, as shown in Chap. 2. Hand programming of these machines is quite complex, and practically all this is done with the use of a computer.

Actually, many of the N/C machines today use only the three major axes. Therefore this book considers only the *X*, *Y*, and *Z* axes. In fact, to simplify the learning of the basic principles of N/C programming, the first few programs discussed omit consideration of the *Z* axis.

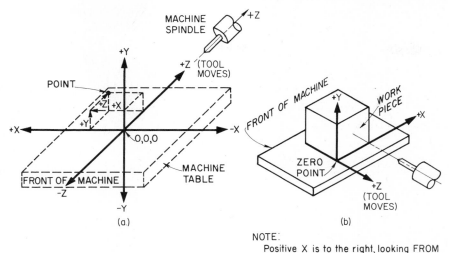

NOTE:
Positive X is to the right, looking <u>FROM</u>
<u>the</u> <u>SPINDLE</u> towards the workpiece.
Positive Z moves <u>tool</u> away from work.

Fig. 4-5   Two ways of showing the standard *X*-, *Y*-, and *Z*-axis notation for horizontal-spindle N/C machines.

The Zero-Point Location

As mentioned previously, the **zero point** is the point where the X, Y, and Z axes intersect, and it is the point from which all coordinate dimensions are measured. This point may be **fixed** by the manufacturer or, as is shown later, in some machines it may be determined by the programmer.

The **coordinate** dimensions (usually called **absolute** dimensions) of any point in Figs. 4-1, 4-4, and 4-5 are always given from the intersection of the axes—the zero point. Thus the dimensions from this zero point to the workpiece must be known. Unfortunately, the location of the zero, even in vertical-spindle machines, is not standardized. Many machines are built with a **fixed zero point.** That is, $X = 0$, $Y = 0$ is at a specific point on the machine table, and cannot be changed (though it can be electronically shifted, as is shown later).

Two examples of this fixed zero are shown in Figs. 4-6 and 4-7. Notice that in Fig. 4-6 all points on the machine table are in the first quadrant. Thus all coordinates will be $+X$ and $+Y$ numbers. However, in Fig. 4-7 all points are in the **third** quadrant (refer to Fig. 4-1), and thus all points on the table and the workpiece will be $-X$ and $-Y$ numbers.

In some machines which have the fixed zero in locations such as that

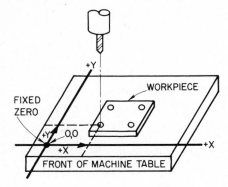

Fig. 4-6 Illustrating a fixed-zero location which results in first-quadrant N/C dimensioning.

shown in Fig. 4-7, the MCU is wired so that the minus signs are not needed. That is, the coordinate dimensions are along only one X and one Y axis, and the machine control unit does not require the use of plus or minus signs. It interprets all dimensions as being along the proper axis and from the established zero point.

On some N/C machines the programmer may select the location of the zero point at any convenient spot. This is referred to as a **floating zero** and is discussed in Chap. 5. There are other variations in this zero-point location, but once the basic programming methods are learned, it is not difficult to handle those situations as they occur.

Setup Point

The **setup point is a point which is actually on the workpiece,** or the fixture holding the workpiece. This point may be the intersection of two previously finished edges, or the center of a previously machined hole in the

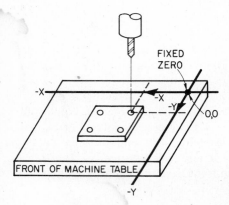

Fig. 4-7 Illustrating a fixed-zero location which results in third-quadrant N/C dimensioning.

workpiece, or a dowel or hole which is at a known location on the fixture. This setup point must be accurately located **in relation to the zero point,** as shown in Fig. 4-8a, and the center of the spindle must be accurately located at the setup point.

Setup procedure in numerical control is much the same as on any machine tool. The part or the fixture must be "lined up" with the table travel and clamped securely in place. Then the center of the spindle must be located in relation to the workpiece, or setup point. These steps are done on N/C machines with dial indicators, wigglers, locating plugs, etc., as on conventional machines, except that the distance from the setup point to the zero point must be held accurately, or be accurate within the capabilities of the ZERO SHIFT ability of the machine (Appendix C).

These special setup locations are shown on a layout of the N/C machine table and the fixture. The tool designer, the N/C programmer, and the setup person all key their drawings and their work to these special locating dimensions. Thus each person's work is coordinated with the work of the other members of the N/C team. Occasionally, as shown later, the zero point and the setup point can be at the same place.

Locating the Setup Point (Fixed-Zero N/C Machine)

In Fig. 4-8 a workpiece is shown with its lower left-hand corner located at (250, 125) from the zero point. This dimension has been decided by the programmer. The location is usually chosen for ease of loading and unloading the part and to keep the part in a convenient place on the machine table. Therefore the programmer would give instructions to the machine operator or setup person to locate the dowel pins (or a fixture) so that when the workpiece was "banked" on these dowels, it would be located as shown, with edges parallel to the motions of the machine table.

Locating the work at the point (250, 125) is complicated by the fact that the fixed zero on numerical control machines is actually an electronic zero point. That is, it is the point from which the electronic circuits measure the dimensions punched into the tape. This point may or may not be actually shown on the machine table. Therefore each N/C machine is equipped with some method of measuring to the setup point. It may require setting dials by hand, or using special *zero set* buttons on the control panel, or punching dimensions into the tape, or a combination of these means for getting started at the right location. These methods are quite simple, and the operator, or setup person, easily learns the method for each make of machine.

It is very important to keep clearly in mind the definitions of the setup and zero points, the use of each of them, and their relationship to each other.

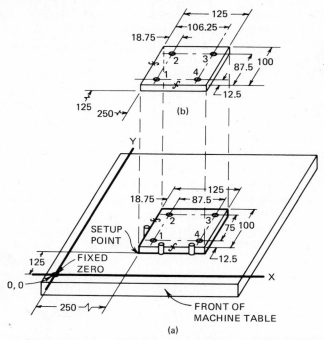

Fig. 4-8   Two methods of dimensioning a workpiece for figuring absolute dimensions from a fixed zero. Dimensions are in millimetres.

In summary,

1. The zero point is the point from which the programmer locates all other points when using coordinate dimensioning.
2. The setup point tells the setup person where to place the fixture or part on the machine table so that the holes and other machining will be in the correct locations on the part when the tape is used.

Now that the basic framework is known, the actual dimensioning of points for N/C work is not difficult. Two methods of dimensioning are used in this work.

Incremental (Delta) Dimensions

In Fig. 4-8a, the distance from the left edge of the workpiece to hole 2 is given as 18.75 mm. From hole 2 to hole 3, the dimension is shown as 87.5 mm on the X axis. Similar dimensioning is given between holes 3 and 4. This is known as **incremental,** or **delta,** dimensioning ("delta" is a Greek letter used in mathematics to signify the "difference" between two quantities). Thus each incremental dimension is given **from the last position** of

Table 4-2 INCREMENTAL (DELTA) DIMENSIONING (FIG. 4-8a)

	Clockwise (CW)			Counterclockwise (CCW)	
	X, mm	Y, mm		X, mm	Y, mm
Hole 1	+268.75	+137.50	Hole 1	+268.75	+137.50
Hole 1	0	+ 75.00	Hole 4	+ 87.50	0
Hole 3	+ 87.50	0	Hole 3	0	+ 75.00
Hole 4	0	− 75.00	Hole 2	− 87.50	0

the machine **to the next position** wanted. To compute the incremental (delta) dimensions for the part shown in Fig. 4-8a, it is first necessary to decide the order in which the holes are to be drilled. This is because, in incremental dimensioning, it is as if each location became a zero point for the next location. All we list is the distance **between** holes, and indicate the **direction** of motion by plus and minus signs. In this type of dimensioning it is as if a **new set of X and Y axes is drawn through each hole,** and the plus or minus signs for the next location follow the rules shown in Fig. 4-1. Thus, if a set of axes is drawn through hole 3, hole 4 is "below" hole 3, and the Y increment is − 75 mm.

Consider this simple part (Fig. 4-8a) if the holes are drilled in clockwise (CW) order 1, 2, 3, 4 and if they are drilled in counterclockwise (CCW) order 1, 4, 3, 2. The result would be as shown in Table 4-2.

Notice that each pair of coordinates is the distance **between** two locations. Thus, in the CW listing, hole 2 is at the same X location as hole 1, so that the X increment is zero, and hole 2 is 75 mm "above" hole 1, so that the incremental dimension is 75.00. It is not usually necessary to actually punch the zero dimensions into the tape, since the machine will move only if given a dimension.

Incremental dimensioning is used on N/C machines with contouring controls more often than on point-to-point machines. Most of the programs in this book are written in the absolute system of dimensioning. Examples of incremental dimensioning are given in Chaps. 8 and 13.

Absolute (Coordinate) Dimensioning

Many numerical control machines require the use of **absolute** dimensions, also known as **coordinate** dimensions. That is, all locations must be given as distances from the **same zero location.** This section shows how coordinate, or absolute, dimensions are figured.

To figure the absolute X and Y dimensions from the fixed zero point to the four holes shown in Fig. 4-8a requires adding dimensions as follows:

Hole 1:

$X = 250 + 18.75 = 268.75$ mm
$Y = 125 + 12.50 = 137.50$ mm

Hole 2:

$X =$ same as hole 1 $= 268.75$ mm
$Y = 125 + 12.50 + 75 = 212.50$ mm

Hole 3:

$X = 250 + 18.75 + 87.50 = 356.25$ mm
$Y =$ same as hole 2 $= 212.50$ mm

Hole 4:

$X =$ same as hole 3 $= 356.25$ mm
$Y =$ same as hole 1 $= 137.50$ mm

These absolute dimensions, as put on tape with decimal points omitted, would be as shown in Table 4-3.

In actual practice it is not necessary to repeat a dimension which is unchanged, such as the 268.75 for hole 2, which is the same as the previous hole, hole 1. This is explained later. Also, the plus signs are usually omitted, as most N/C machines assume that all numbers are positive unless a negative sign is used.

Notice that in absolute dimensioning, the coordinates of a point are the same, no matter in which order the points are listed. Notice also that "zero" (0.0, 0.0) is, in coordinate dimensioning, an actual point location, not a "difference," as in incremental dimensioning.

By now you may have realized that there are no fractions or decimal equivalents to be learned when using metric dimensions, which is one advantage of this system.

Baseline Dimensioning of the Drawing

Figure 4-8b shows the same part drawn using baseline dimensioning. This type of dimensioning makes it much easier for the N/C programmer.

For example, using the drawing of Fig. 4-8b:
Hole 2:

$X = 250 + 18.75 = 268.75$ mm
$Y = 125 + 87.50 = 212.50$ mm

Hole 3:

$X = 250 + 106.25 = 356.25$ mm
$Y =$ same as hole 2 $= 212.50$ mm

Table 4-3 COORDINATE
DIMENSIONS OF HOLES IN
FIG. 4-8*b*

	X, mm	Y, mm
Hole 1	+268.75	+137.50
Hole 2	+268.75	+212.50
Hole 3	+356.25	+212.50
Hole 4	+356.25	+137.50

The time saved and the decreased chance of error in this simple example are small. However, when locating tens of holes and surfaces, the saving can be considerable. Some companies, when making new drawings, plan ahead, and if N/C is to be used to machine the part, the draftsmen will seriously consider using baseline dimensioning. In fact, some companies prefer this type of dimensioning for most of their drawings.

There are several ways of showing baseline dimensioning, and the problems in this book are dimensioned in a variety of methods in order to illustrate the different systems. Many systems of dimensioning are in use, and justifiably so. Whatever system is used, the programmer for numerical control machines must work accurately and methodically in figuring all dimensions. Haste and undue use of shortcuts often cause errors—and errors cost money.

Chapter 5

TAB SEQUENTIAL PROGRAMMING

Before discussing the details of hand programming, it is necessary to understand what happens when a punched tape is "read" into a numerical control machine.

From Tape Code to Machine Action

The electrical and electronic equipment which reads the punched-tape codes and changes them into lighted signals, machine motion, speeds, etc., is made up of several units which work together in a four-step process, as shown in Fig. 5-1.

1. The tape reader (Chap. 3 and Appendix D) **reads** the holes in the tape and changes these into electrical signals which are sent to the central control of the system. This center is called the machine control unit (MCU), or sometimes just the controller, or the director.

2. The machine control unit (MCU) is, in some ways, similar to a small computer. It is made up of easily replaced solid-state control circuits which **change the signals** received from the tape reader into signals to the numerical control machine.

   The MCU "remembers" some of the commands; so they do not need to be repeated. It sends the tape commands to the proper motors and lights on the machine and to the control console, and it controls machine motions, described next.

3. The numerical control machine moves (using electrical or hydraulic power) and, in most N/C machines, **generates a feedback**

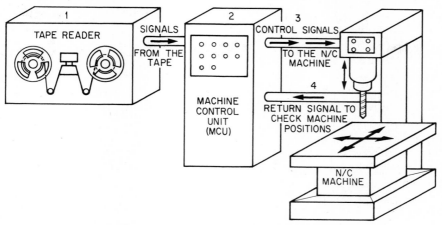

Fig. 5-1 The process of converting tape codes into machine action.

    signal telling the MCU how far it has actually moved along each tape-controlled axis.

4. The machine control unit **compares** the machine's actual locations during the movements with the distances or locations originally called for by the tape signals. When all the machine locations agree with the tape signals, the machine motions are stopped, and the next block of information is read into the MCU.

Some N/C machine builders make their own control units, but many of them buy these units from one of several manufacturers who specialize in this type of equipment. In either case, the MCU can have built into it whatever capabilities the buyer wants. There are many optional choices, such as:

    Simple point-to-point
    Added ability to cut circles (circular interpolation)
    Contouring control (circular and linear interpolation)
    Two- to five-axis control
    Readout of X, Y, and Z locations
    Milling controls
    Feed rate control
    Speed control

Of course, the extra controls are more expensive; so each shop orders equipment to meet its particular needs and budget. The machine-tool builder includes the necessary mechanical, electrical, and hydraulic equipment to give machining and machine movement capabilities to match the control system.

How Tab Sequential Works

On a standard typewriter used for office work, there is a tabulation key. This is usually abbreviated "tab" on the typewriter key. When the tab key is pressed, the typewriter carriage moves rapidly to a preset stop. Thus the typist can easily keep the data arranged in neat columns on the typed record, as shown in Fig. 5-2.

Similarly, when making an N/C tape, it is easier to check the figures if all the X, Y, and other values punched on the tape are arranged in columns on the typed record, as shown in Fig. 5-3. When used during the making of an N/C tape, the TAB key on a tape punch does two things at the same time: it spaces the columns on the typeout and also punches the five-hole tab symbol in the tape (Fig. 3-8).

In TAB SEQUENTIAL programming, this five-hole tab symbol is fed from the tape, through the tape reader, and into the machine control unit (MCU). The electronic circuitry of the MCU interprets this tab punch as a signal that the next set of numbers which is going to be read from the tape must be fed to a different section of the machine control system. In this

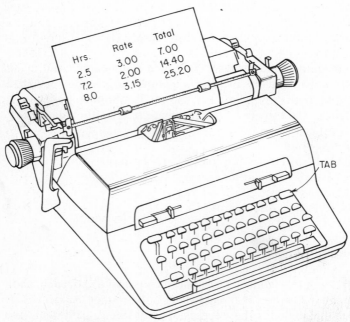

Fig. 5-2   Tabulated columns in office typing.

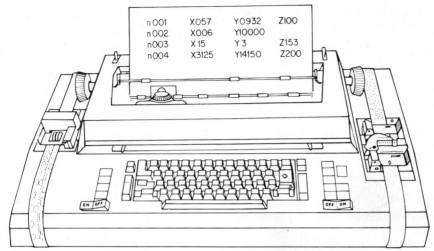

nOOI	XO57	YO932	ZIOO
nOO2	XOO6	YIOOOO	
nOO3	X I5	Y 3	ZI53
nOO4	X3125	YI4I5O	Z2OO

Fig. 5-3 Tabulated columns in numerical control.

system of programming, the tab code is thus the dividing line between signals to different parts of the N/C machine.

This action is similar to what happens when an electromechanical device, called a **stepping switch,** is used. This is simply a switch which can be rotated to connect with any one of a number of separate contacts. Each contact on the switch feeds the common input current to a different part of a machine or through a different electric circuit.

To illustrate how this switching works we assume that a single-vertical-spindle two-axis N/C machine is being used. The information which must be punched into the tape in order to program hole 1 in Fig. 5-4 is:

010   SEQUENCE NUMBER. The line, or operation number, of the program (for reference only).

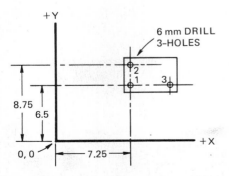

Fig. 5-4 Hole locations programmed in Figs. 5-5 through 5-8.

Table and carriage are to move so that the center of the spindle is located at

$X = +7.250$ mm   from the zero point
$Y = +6.500$ mm   from the zero point

06   TOOL CHANGE. The machine (and the tape) are to stop so that the operator can change the tool. This is called the M, or MISCELLANEOUS, column.

EOB   END OF BLOCK. This is all the information needed to drill at location 1.

Since no decimal points are punched in an N/C tape, the actual tape typeout would look like this:

010 07250 06500 06

Notice that the TAB and END OF BLOCK signals do not show on the typeout, since they only punch holes into the tape and move the typewriter to different positions. Notice also that the plus signs are omitted, because this N/C machine calls all numbers positive unless they have a minus sign in front of them.

In Fig. 5-5a to e, each rectangle at the top of the drawings represents a section of the machine control unit which will receive and send signals for **only one** action of the N/C machine. Thus all numbers entering the X section will cause table motion left or right on the X axis, and all signals entering the Y section will cause motion in or out on the Y axis. The SEQ NUMBERS may show in lights on the console, and the MISC signals will perform special jobs.

Figure 5-5a

When the tape starts through the tape reader, the switch is in the extreme left position, connected to the SEQUENCE NUMBER section. Thus the first numbers 010 will be sent into the SEQ and will, if a SEQUENCE READOUT is on the machine, light up numbers on the panel to help the operator know which part of the work is being done.

Figure 5-5b

Now the first TAB signal is read. This causes the switch to move one step to the right and connect to the X control circuit. Then 07250 is read into this section of the control.

Figure 5-5c

The second TAB moves the switch so that it connects to the Y control section, and the Y dimension, 06500, is read into the MCU.

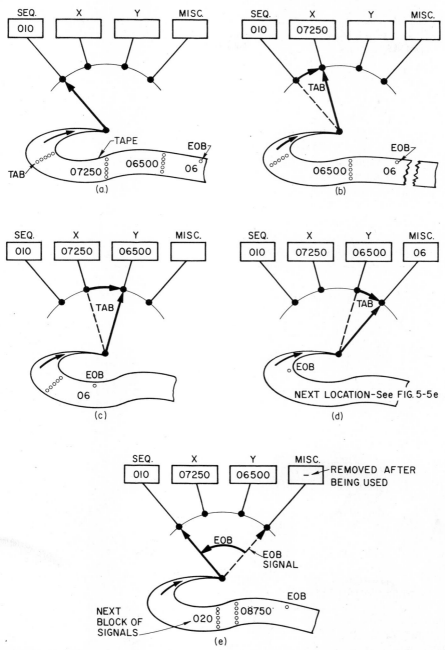

Figure 5-5 Schematic drawing of the action of the MCU when receiving the information for drilling hole number 1, Fig. 5-4. TAB SEQUENTIAL format.

Figure 5-5d

The third TAB moves the switch to the MISC section, and the 06 code is read from the tape.

Figure 5-5e

The END OF BLOCK signal tells the machine, "That's all, now go ahead and perform the operation." At the same time the switch returns to the left, ready to read the next block of information. The tape is stopped until the operation is completed.

It is important to notice that the MISC function 06 is removed at this time, **but all the other numbers stay in the MCU's memory.** Not all MISC functions are removed by the EOB. This is explained in later chapters.

All of this happens at a rate of from twenty to several hundred rows (or characters) per second. Thus, at a rate of sixty rows per second, it would take $^{19}/_{60}$ sec = 0.32 sec to read this whole block of information.

Figure 5-6 shows all these steps combined in one diagram.

Handling the next set of signals is done as illustrated in Fig. 5-7, which

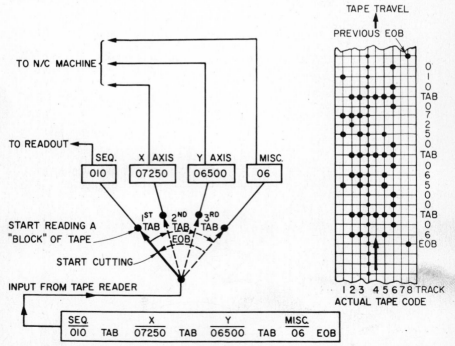

Fig. 5-6 Complete stepping-switch diagram and the punched-tape coding for TAB SEQUENTIAL format.

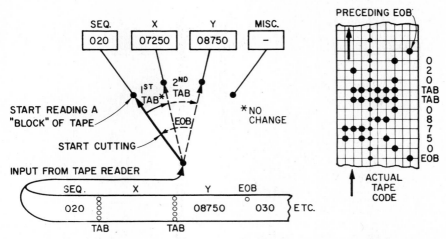

Fig. 5-7 Second block of tape and complete stepping-switch diagram.

shows, at the bottom, the new set of coordinates for drilling hole 2 of Fig. 5-4. The numbers in the rectangles at the top of Fig. 5-7 show the final result after reading the section of tape which has these figures coded in it.

Figure 5-8*a* to *c* shows how this was done.

Figure 5-8a

Notice that the previous *X* and *Y* values are still in the MCU's memory.

Figure 5-8b

The first tab code moves the switch to the *X* block, but no signal is needed because the *X* coordinate does not change.

Figure 5-8c

The second tab moves the switch to the *Y* block and feeds the 08750 value into the MCU.

No signal is needed in the MISC block because we are still using the same drill.

Notice that if a change in the *X* dimension only is required (as in moving from hole 1 to hole 3 in Fig. 5-4), it would only be necessary to list

1. SEQUENCE NUMBER
2. TAB
3. *X* value (no *Y* value because *Y* does not change)
4. EOB

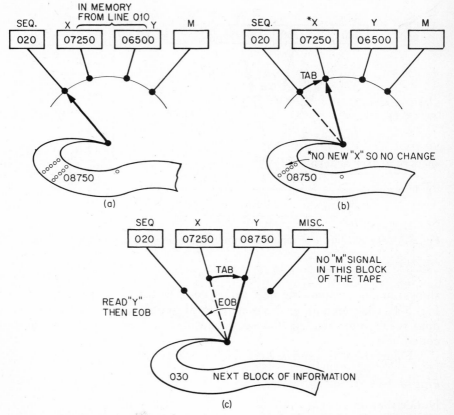

Fig. 5-8 Schematic drawings of the action of the MCU when receiving the information for drilling hole number 2 (second block in the tape). Fig. 5-4. TAB SEQUENTIAL format.

Of course, the electronic equipment is very much more complex than this description. However, the resulting process may be visualized as described, and it will help in understanding why we can take certain shortcuts in programming.

Full Floating Zero

Some N/C machines, such as the one shown in Fig. 5-9, do not have a fixed zero point. These have an MCU which allows the programmer to select any point for the zero and then compute the programmed dimensions from this point. It is as if you picked up the X, Y coordinate system and put it down on the workpiece with the (0, 0) intersection wherever you

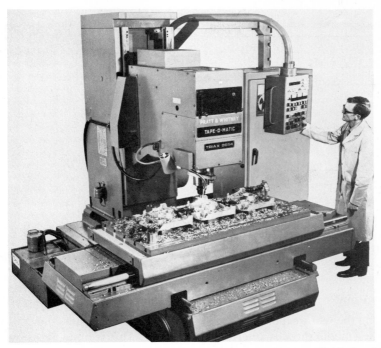

Fig. 5-9 A modern two- or three-axis N/C machine. Available in either inch or metric. Capacity, 50-mm drill, and up to 45 cm³/min milling. X and Y axes accurate to ±0.02 mm or better. (*Colt Industries, Pratt & Whitney Machine Tool Division.*)

thought it would make your programming work the easiest. Thus the workpiece may be clamped down and lined up at any convenient place on the machine table.

In this system you must now locate a **setup point** at a location which is an accurately known distance from our selected zero point. This setup point may be a corner of the workpiece (if the edges are finished), or a previously bored hole, or the finished I.D. or O.D. of a previously machined hub. It can even be a specially prepared location on the holding fixture.

In Fig. 5-10 the dimensions are mostly from the centerlines, so the zero point was selected at their intersection. This type of zero location often creates point locations in all four quadrants. Thus it is a good idea to actually draw the X and Y axes on the part print, as shown in Fig. 5-10, preferably with a colored pencil. This will clearly show in which of the four quadrants each hole is located and will help to avoid errors in plus or minus signs.

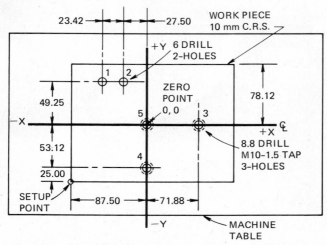

Fig. 5-10   Workpiece for part program shown in Fig. 5-11. FLOATING ZERO, with setup and zero points not at the same location. Dimensions in millimetres.

The Setup Point on the Tape

After aligning the workpiece and securing it, the operator locates the centerline of the spindle directly over the setup point. This is done by using indicators or wigglers or edge-finders, etc., as in any machine-shop work.

But the MCU does not yet "know" where the spindle is located in relation to the (0, 0) point which was selected. Thus the first information on the tape must, in this type of machine, be the location in X and Y of the setup point **from the zero point.** Remember, the (0, 0) is the reference for **all** locations, including the setup point.

In Fig. 5-10 the centerline of the spindle has been "zeroed in" to the lower left-hand corner of the workpiece, and this is in the third quadrant. Thus the coordinates of the setup point are $X = -87.50$, $Y = -78.12$. These, for this type of machine, must be the first numbers punched into the tape. Naturally, no machining is to be done at this point. However, these two coordinates must be read into the MCU, and then, by pressing the "zero-set" buttons, we shall have dimensioned the setup point, thus giving the MCU a starting point for all future dimensions.

Writing a Program—Tab Sequential—Floating Zero

The actual writing of a program can now be shown. The following program refers to machining the drilled and tapped holes in Fig. 5-10 on a single-vertical-spindle N/C machine with no attachments except a box showing the sequence numbers for the operator's convenience. Leading zeros will be omitted. The machine can handle three digits to the left of the decimal point and two digits to the right. Thus the last digit to the right will be read as 0.01 to 0.09 mm.

The codes to be used are:

02 Stop the machine; rewind the tape.

06 Stop the machine and the tape so that the operator can change the cutting tool. After changing the tool, the operator pushes the *start* button. This advances the tape to read the next block and continue machining.

EOR Rewind stop (End of Record).

Vertical (Z axis) travel is automatic to stops which have to be hand-set by the operator.

Writing the N/C program is usually done on preprinted programming sheets. The actual arrangement of the sheets for different machines will vary according to the amount of information needed on the tape. However, all of the forms have much in common. Having once used three or four basic types of programming sheets, you will not have difficulty with a slightly different arrangement.

The form shown in Fig. 5-11 is one of the simplest. The meanings of the column headings are as follows:

SEQ. NO. The sequence number is merely a convenient way of identifying each line in a program. This can consist of one, two, or three digits, usually three. These numbers can be 1, 2, 3, 4, etc., or 5, 10, 15, 20, etc., or 10, 20, 30, 40, etc.

For hand programming these lines are often numbered by 5s or 10s (as in many process sheets, or operation sheets, used in production). This allows extra operations to be easily added if needed. For example, if it is necessary to add a DRILL AND REAM or COUNTERSINK TO BURR when a simple DRILL was originally planned, the sequence numbers 2, 6, 11, 14, etc., could be used.

Frequently, these sequence, or line, numbers are shown in lighted numbers on the console of the MCU to help the operator to follow the instruction sheet.

TAB or EOB "Tab," of course, refers to the tab punch, which is required in TAB SEQUENTIAL programming. EOB, as mentioned earlier, is END OF BLOCK. It is sometimes abbreviated EB, or just E, or occasionally EL or EOL, meaning END OF LINE. The TAB and EOB columns are frequently omitted on forms used by experienced programmers since they write in the EB wherever it is needed, and the tape typist is trained to tabulate all columns in a TAB SEQUENTIAL program.

±X and ±Y The ± sign is a reminder that some programs require signed numbers. This reminder is omitted in many N/C forms that are used in industry.

MISC.	SEQ. NO.	TAB EOB	± X	TAB EOB	± Y	TAB EOB	M FUNCT.	EOB	REMARKS†
	000	T	-87|50	T	-78|12	EB			SET UP INFORMATION. READ TAPE, THEN PUSH "ZERO" BUTTONS
EOR*%	010	T	-50|92	T	+49|25	EB			6mm DRILL - HOLE #1 1500 rpm
* &	020	T	-27|50	T	—|—	T	06	EB	6mm DRILL-HOLE#2 CHANGE TOOL
	030	T	0|	T	0|	EB			8.8mm DRILL-HOLE#5
	040	T	+7|88	EB					-HOLE#3
* &	050	T	0|	T	-53|12	T	06	EB	-HOLE#4 / CHANGE TOOL
	060	T	EOB						M10-1.5 TAP-HOLE#4
	070	T		T	0|	EB			-HOLE#5
	080	T	+7|88	EB					-HOLE#3
*/ &	090	T	+120|00	T	+95|00	T	02	EB	MOVE SPINDLE OFF PART. CHANGE TOOL. TAPE REWINDS.

*THESE SYMBOLS ARE USED ON THE OLDER TAPE-O-MATIC MACHINES INSTEAD OF THE EOR AND M FUNCTION CODES.
†REMARKS ARE NOT PUNCHED INTO THE TAPE.
% = REWIND STOP, & = TOOL CHANGE, / = TAPE REWIND

Fig. 5-11   The N/C program for drilling and tapping Fig. 5-9. TAB SEQUENTIAL, with FLOATING ZERO. Note that both plus and minus absolute coordinates are required. Leading zeros are omitted.

MISC   This column is for the MISCELLANEOUS instructions to the machine. As will be seen in later chapters, these instructions can be quite varied. They are always coded with **two** numbers, one or both of which may be zero. In this first program, only 02 and 06 are used, as explained in preceding paragraphs.

REMARKS   As shown in Fig. 5-11, the "remarks" column contains important instructions to the operator. In this simple program, rpm (spindle speed) and hand-set feed are shown. On more sophisticated machines, these are controlled by the tape. Remarks are not punched into the tape.

Tool specifications or tool numbers may also be shown as instructions to the operator. Only in quite expensive N/C machines are the tools automatically changed by tape commands.

Usually, a duplicate of this programming sheet or a copy of the typeout is made. This copy is sent to the N/C machine operator with the operation sheet, the tooling, the parts to be machined, and the tape.

Explanation of the Program of Fig. 5-11, Part Drawing Fig. 5-10

Note: All dimensions are in the absolute (coordinate) system, and in millimetres (mm).

Seq. 000 Setup point dimensions.
$X = -87.50$, as it is in the third quadrant.
$Y = -53.12 - 25.00 = -78.12$.
No MISC or M functions used.
Cutting speed = 30m/min (Appendix E).
rpm = $300 \times 30 \div 6 = 1500$.

Seq. 010 Hole 1
$X = -27.50 - 23.42 = -50.92$.
$Y = +49.25$.
EOR (End of Record) or similar punch is used in the MISC column to stop the tape when it rewinds, as there is no machining at Seq. 000, the setup point.

Seq. 020 Hole 2
$X = -27.50$, directly from the drawing.
Y = same as hole 1, so two tabs in succession are punched and the +49.25 stays in the machine's memory.
06 = Tool-change light comes on.

After drilling hole 2, the tape and the machine stop so the operator can change to an 8.8 mm drill. Push *start* to continue.

Seq. 030 Hole 5

Note: This could have been hole 3 or 4, but it is good practice to keep machine travel to a minimum, and hole 5 is the closest to the last hole drilled.

$X = 0$, as zero point was selected here.
$Y = 0$, also.
No other information needed.

Seq. 040 Hole 3

Note: Could have gone to hole 4 instead.

$X = +71.88$, directly from the drawing.
EB after the X value as no other changes are needed.

Seq. 050 Hole 4
$X = 0$.
$Y = -53.12$, directly from the drawing.
06 Tool-change light goes on. Tape and machine stop. Operator in-

stalls a clutch-type reversing tapping head with an M-10 tap. The depth stop may also be adjusted. Pushes *start* to continue.

Seq. 060   Hole 4

Note: The first hole tapped will be hole 4 as the spindle is already there. There is no change in either *X* or *Y* so the machine reads the EB and taps hole 4. Using this simple machine, the operator might hand-feed the tapping operation.

Seq. 070   Hole 5

Note: Could have gone to hole 3.

X = still at zero, so tab past it.
Y = 0, same as in line 030.
EB follows *Y* coordinate.

Seq. 080   Hole 3

X = +71.88, same as in line 040.
Y = unchanged, so EB follows the *X* coordinate.

Seq. 090

Note: This extra positioning is often used to get the spindle out of the way so that the workpiece can easily be removed.

X = +120.00 moves the table to the left so that the spindle is to the right of the part. (An *X* of −120.00 would leave the spindle to the left of the work.)
Y = 95.00 moves the table forward so that the spindle is in back of the work.
02   Stop machine. Tape rewinds to the "rewind stop" punch.

The operator removes the finished part, secures the next piece, and changes the tool to a 6-mm drill.

Operator then pushes the *start* button and the MCU reads block 010 and begins a new cycle.

Notes: No tolerances are shown for these dimensions because the tolerance will be whatever the N/C machine will produce. However, tolerances are frequently shown on the original drawing so that the inspection department has a guide. Errors do occasionally occur in tapes and readers, and chips do cause misalignment of parts.

Notice that zero is **not** punched into the tape to represent a blank space. *X* = 0, *Y* = 0 is a definite location on all N/C machines. Thus zero is not punched into the tape unless the spindle is to be located at that point.

On a FLOATING ZERO machine, as in this program, the (0, 0) point is frequently (but not always) used.

Another Zero-Point Location

In Fig. 5-12 the previously programmed part is shown with two changes made.

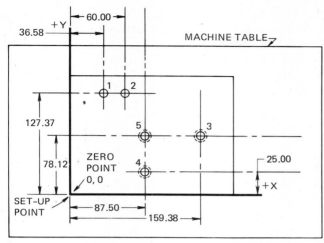

Fig. 5-12 A baseline dimensioned drawing showing how, on a FLOATING ZERO N/C machine, the zero point and the setup point may be at the same place. Dimensions in millimetres.

First, the zero point has been located at the same place as the setup point. In this arrangement all hole locations are in the first quadrant, so no negative values will be used.

Second, baseline dimensioning, starting at zero, has been used. This makes programming very easy, as all coordinates can be read directly from the drawing.

With this arrangement, the first line of the program will be $X = 0$, $Y = 0$ since there is no difference in location between the setup point and the zero point. Notice that in this case no machining will be done at the (0, 0) location.

The first few lines of this program are:

000 T 0 T 0 EB (Push zero-set buttons)
EOR 01 0 T 3658 T 12737 EB (Hole 1)
020 T 6000 T T 06 EB (Hole 2—change tool)
030 T 8750 T 7812 EB (Hole 5)
040 T 15938 EB (Hole 3)
etc.

In full floating zero programming it is important to remember the following:

    1. Sequence number 000 is never used to specify a hole location for machining. It is only the setup dimension. If a hole is needed at this location, it must be programmed in a separate block.

2. The EOR is punched before the first block of information which actually specifies a machining location. There is no TAB between the EOR and the 010. When the tape rewinds, after the 02 code, it will stop when it comes to the EOR. Thus, sequence number 000 will be read for only the first part machined. The setup information is held in the MCU's memory until the machine is shut off, or until a new set of setup dimensions is put into memory by use of the zero buttons.

3. The plus and minus signs must be properly assigned to each coordinate. The plus signs, of course, do not have to be punched into the tape. However, it may be wise to show them on your first few programs, as a reminder that signed numbers are required.

4. Note that since leading zeros are omitted, only one zero is needed to call for a zero location in the X or Y axis. Up to five zeros may be used, but there is no advantage in doing this.

A Fixed Zero–Tab Sequential Program

Many N/C machines have a FIXED ZERO, often at one of the four corners of the machine table. This is an electronic zero—where the spindle will locate if (0, 0) is called for on the tape. This is almost never a machining location.

The zero in Fig. 5-13 is at the lower left-hand corner, which makes all coordinates plus, so no signed numbers are needed.

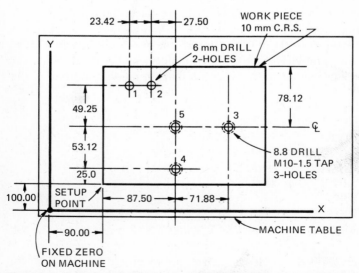

Fig. 5-13 Sample workpiece, located from a fixed zero, dimensioned from the centerline of one of the holes (not the best way). Dimensions in millimetres.

MISC.	SEQ. NO.	TAB EOB	±X	TAB EOB	±Y	TAB EOB	M FUNCT.	EOB	REMARKS
SETUP AT X=90.00, Y=100.00 - LOWER LEFT CORNER									
EOR	005	T	126 58	T	227 37	EB			HOLE 1 = 6mm DRILL
	010	T	150 00	T	——	T	06	EB	HOLE 2 - CHANGE TOOL
	015	T	177 50	T	178 12	EB			HOLE 5 - 8.8mm DRILL
	020	T	249 38	EB					HOLE 3
	025	T	177 50	T	125 00	T	06	EB	HOLE 4 - CHANGE TOOL
	030	T	EOB						HOLE 4 - MIOTAP
	035	T	——	T	178 12	EB			HOLE 5
	040	T	249 38	EB					HOLE 3
	045	T	290 00	T	300 00	T	06	EB	STOP-REWIND TAPE CHANGE TOOL-OFF THE WORK.

Fig. 5-14 A TAB SEQUENTIAL program from a fixed zero, for drilling and tapping.

The codes and N/C machine used for this program are the same as those previously cited, except for the fixed zero.

Setup procedure is to accurately position and align the part on the machine table so that the lower left-hand corner of the workpiece is at X = 90.00, Y = 100.00 mm. This the operator can do by hand, dialing these numbers on the MCU console (see Fig. 6-3 and 6-4). This moves the table to the setup position, and then the operator indicates the two edges and clamps the part to the table.

Of course, the workpiece will be raised above the table on parallels or clamped to a fixture so that the drills can go through the workpiece and not drill into the machine table.

The setup dimensions of 90.00 and 100.00 mm are purely arbitrary. They are decided by the programmer and may be different for different size work and larger or smaller machine tables.

The program may, as in floating zero, begin at any hole location, and the holes may be machined in any order desired. If hole locations must be held to ±0.08 mm [.003 in.] or less, it is usually necessary to first center-drill or spot-drill all hole locations. To save time and repetition, spot drilling has been omitted in all the programs in this book.

Writing the Program—Fixed Zero

With a fixed zero, there is no "zero location" block. The programmer usually notes at the top of his form the setup information for the operator. Thus, the first block of information will be a machining location.

Fig. 5-14 shows the program for the drawing Fig. 5-13. The work is much the same as in Fig. 5-11 except that the $X = 90.00$, $Y = 100.00$ must be added to all the part dimensions. For example:

Seq. 005   Hole 1
$X = 90.00$ (setup) $+ 87.50$ (to center) $- 50.92 = 126.58$
$Y = 100.00$ (setup) $+ 25.00 + 53.12 + 49.25 = 227.37$
Seq. 010   Hole 2
$X = 90.00 + 87.50 - 27.50 = 150.00$
$Y =$ same as hole 1, so TAB past it.
$06 =$ Tool change to 8.8 mm tap drill.

Fig. 5-15   Work being drilled on an N/C machine which uses FLOATING ZERO and TAB SEQUENTIAL coding. Notice the locating blocks at the back left-hand corner and the simple clamping. (*Colt Industries, Pratt & Whitney Machine Tool Division.*)

Notice how much easier it would have been if this part had base-line dimensions from the lower left corner of the workpiece.

The rest of the program should be easily worked out by the student. Notice that the order of tapping holes is not the same as when they were drilled. This makes no difference, though the program is somewhat changed.

The two part-programs described in this chapter are simpler than most found in industry. However, a part with many hole locations merely requires more lines in the program. No additional programming knowledge will be needed if you are using an N/C machine similar to the ones specified in this chapter.

Many numerical control machines have more tape-controlled functions than have been considered so far. The next chapter describes another widely used system of programming and a numerical control machine which has two additional tape-controlled capabilities.

Chapter 6

WORD ADDRESS PROGRAMMING

An **address** is a specific location. In daily use we think of a person's or a company's address having several lines of information. In numerical control, however, the address is a single letter from *A* to *Z*. In fact, WORD ADDRESS programming is sometimes referred to as LETTER ADDRESS programming.

The Electronic Industries Association (EIA) and the Aerospace Industries Association (AIA) have published suggested standards. Parts of these are shown in Appendix F. These standards suggest meanings for the letter and number codes used for N/C programming.

Most N/C machine manufacturers today follow these recommendations quite closely. In fact, machines using TAB SEQUENTIAL programming often use the **number** codes suggested for G and M, but of course without the letters. The 02 and 06 used in Chap. 5 are two of these numbers.

How Word Address Works

In Fig. 6-1, nine "words" are shown. Some machines use fewer, and the contouring (NCC) machines sometimes use several more. If the example shown in Fig. 6-1 were programmed in TAB SEQUENTIAL, it would be:

065T81TTT02550TTTT06   (T means "tab")

This requires the use of eight tab codes to get the numbers into the correct section of the machine control unit (MCU), which is actually not very difficult. However, an error in counting the number of tabs between

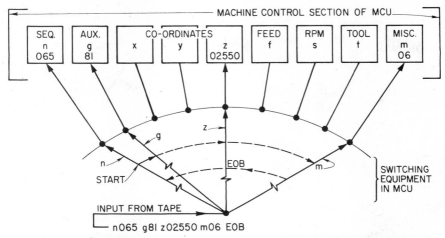

Fig. 6-1 Schematic diagram of WORD ADDRESS programming, illustrating the switching action called for by each letter in the block.

codes would cause an incorrect motion of the machine, or would stop it, because of an error signal.

In the WORD ADDRESS (or LETTER ADDRESS) system, the location to which a signal is to go is identified by the single-letter "word" code instead of by the signal's location in the block of tape, as it is in the TAB SEQUENTIAL system. Figure 6-1 shows how a machine control unit would sort out the information in a block of tape, as if each letter operated a magnet which pulled the switch to the proper location in the MCU electronic circuitry.

Thus, it is not always necessary to arrange the signals in any special order. Of course, the order of each line of a program is usually the same, in order to simplify programming and checking. In the example in Fig. 6-1, the program might be written on a form sheet, with columns for each of the nine functions. However, the typed and taped information would look as shown.

Note: Several N/C machines allow the use of tab even with a WORD ADDRESS program. In these machines the MCU simply disregards the tab code. However, its use allows the information to be typed in a neat arrangement of columns, at the same time that the tape is being punched, if this is of advantage in a company's system.

End of Block (EOB or EB)

In all programming from now on, it will be assumed that EOB is at the end of every line. The tape typist **must** press *carriage return* to start a new

SEQUENCE NUMBER

Fig. 6-2 An example of SEQUENCE NUMBER display.

line. Thus EOB will automatically be punched; so it will not be shown on the programs which follow.

Readout of n, x, and y

Many N/C machines are equipped with a SEQUENCE NUMBER READOUT, and POSITION or TAPE COMMAND READOUTS for the X and Y axes, and sometimes the Z axis also. These are sets of lighted numbers mounted in the front of the MCU or on the operator's console.

The SEQUENCE NUMBER display (usually three digits) identifies the block or line of information being read from the tape (Fig. 6-2). In a semi-automatic N/C machine, such as that specified for use in this problem, the machine operator can refer to the program sheet and select the tools, feeds, and speeds specified for the sequence number displayed.

The position indicators on most NPC machines show the X and Y (and sometimes Z) locations from zero as called for by the tape. Notice that the decimal point is indicated on the panel (Fig. 6-3) and that usually five or six digits are shown, though N/C machines may use more.

Many machines are equipped with manual-input dials (Fig. 6-4), which allow the operator to dial in any coordinates that are needed. Then, with the machine in the manual mode of operation, the table will move to the location which has been selected. This ability is especially valuable, and is sometimes a necessity, during setup, for locating the fixture or

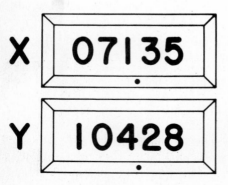

Fig. 6-3 An example of X- and Y-axis TAPE COMMAND READOUT display, for metric dimensions. For inch dimensions, the decimal point is one digit further to the left.

X AXIS

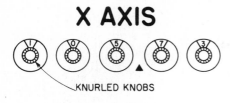

KNURLED KNOBS

Y AXIS

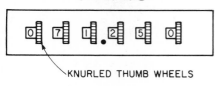

KNURLED THUMB WHEELS

Fig. 6-4 Two styles of manual-input dials. Decimal point shown for reading in millimetres.

workpiece in relation to the zero, or starting point. Also, if engineering changes are made, the machine can be manually operated to locate and machine the new holes until a corrected tape is available. These manual-input dials may be used with any type of programming—TAB SEQUENTIAL, WORD ADDRESS, or any other. Some N/C machines also have manual-input dials for speeds, feeds, and the n, g, and m codes.

The N/C Machine Used in This Chapter

Most N/C machines today have more capabilities than the very simple one we considered in Chap. 5. Thus the N/C machine for use in this chapter's WORD ADDRESS programming has the following capabilities:

1. Single-spindle-vertical, with display of sequence number and X and Y coordinate values.
2. Automatic selection of vertical-spindle motions for drilling, boring, milling, and tapping; sometimes called canned cycles.
3. Depth (Z-axis) control by means of preset cams or stops. These can be "called out" by tape command, but are hand-set by the operator before machining is started.
4. Speeds and feeds must be set by the operator each time a change is required. Maximum speed is 4000 rpm, maximum feed = 760 mm/min.
5. Fixed zero at front left-hand corner of table.

Notice that items 2 and 3 in this list are capabilities requiring taped information which was not included in Chap. 5. The N/C machine shown in Fig. 6-9 has capabilities similar to the foregoing description.

The "Words" Used by This Machine

Note: Either capital or small letters may be used in writing a program since they are the same BCD coding in the tape.

n   SEQUENCE NUMBER. Same meaning as in TAB SEQUENTIAL.

g   PREPARATORY function. This calls on a canned cycle built into the machine, causing the spindle to operate in a specified cycle. See the list of g commands in the next section of this chapter.

m   MISCELLANEOUS functions. These are commands, such as tool change, cam selections, etc. See list of m commands following the list of g commands.

x   *X* coordinate dimension.

y   *Y* coordinate dimension.

List of g (Preparatory) Functions

The list of PREPARATORY functions for point-to-point programming (NPC) by EIA assigns definite meaning to only a few numbers. The following list is compatible with EIA and similar to the numbering used by many N/C machines. (g78 and g79 are "unassigned" in the above codes.)

g78   MILL cycle STOP. The table or spindle first moves at rapid traverse to the programmed *X* and *Y* positions. The spindle (or more accurately the quill) then moves down at rapid advance down to the clearance line (sometimes called the feed engagement point), which was previously set on the cam control. From this depth the cutter is fed down at the previously set feed rate to the final depth set on the cam control.

   The cycle then stops, and the clamp-quill light is lit. The operator clamps (or unclamps) the quill and pushes the *cycle start* button to continue the job.

   This command is used before and after a milling operation which requires that the quill be clamped in order to hold a depth more accurately or to hold a specified dimension. Today this code is less often used, as newer machines automatically clamp and unclamp the quill.

g79   MILL cycle. After this command, the spindle first locates at the specified *X* and *Y* locations at **feed rate.** It then goes to the cam-set depths at rapid and feed rates, and stays down. The next tape command is immediately read. This may be a move in *X* or *Y* for the purpose of a milling cut. This command is used either by itself, if it is not necessary to clamp the quill, or after a g78, to start the actual milling cut in *X* or *Y*.

g80   CANCEL cycle. This command causes the machine to move at rapid traverse to *X* and *Y* as programmed, but the spindle does not move down at all. The next block of tape is read immediately. This is used to keep the spindle and cutter up out of the way when moving over clamps and projections on the workpiece. It is also used to position rapidly before a following g79 milling command.

g81   **DRILL** cycle. The table (or spindle) first moves at rapid traverse to the programmed X and Y positions. The spindle (quill) then moves rapidly down to the clearance plane (or feed engagement point), which was previously set on the cam. From this depth, the cutter is fed downward at the preset feed rate. When it reaches the final depth set on the cam, the feed will stop, and the cutter is retracted at the rapid rate. The next block of tape is then read, and the work is continued.

This command is also used for countersinking, counterboring, spot drilling, and spot facing. When a g81 is used after a g78 or g79 milling command, it raises the spindle to home position at rapid travel.

g84   **TAPPING** cycle. Machine positions at rapid traverse in X and Y according to the tape commands. The spindle moves down at rapid, and feeds to the cam-controlled depth at feed rate. Spindle rotation is then reversed, and the spindle rises at feed rate until the tap is above the feed engagement point. The spindle then continues upward at rapid travel, and rotation is changed to the standard CW direction.

g85   **BORE** cycle. This is the same as g84, except that the spindle does not reverse. This brings the boring tool up at feed rate, which avoids marking the finish. For rough boring, if a spiral mark is no disadvantage, a g81 command may be used for boring.

Note: Not all numerical control machines use the codes described above. The milling command codes vary especially, since they are not in the EIA Standards. However, these are the definitions used in this chapter and the next chapter.

List of m (Miscellaneous) Functions

This list is in accordance with EIA coding. The first four codes are used (with and without m) by many N/C machines. (m50 to m99 are **unassigned** in the codes.)

m00   **PROGRAMMED STOP.** This code stops the machine and the tape at the end of the programmed cycle for inspection or adjustment, and the spindle retracts. Push *cycle start* to continue program. Spindle and coolant may stay on in some machines or may stop, according to the machine.

m02   **END OF PROGRAM.** Spindle stops and coolant is off, after completing the commands in the block. The tape rewinds or a "looped" tape advances to the start of the first command, whichever method is selected by a console switch. All functions are cleared; nothing is left in the MCU's memory.

m06   TOOL CHANGE. This is the same as m00, but tool-change light goes on, and spindle rotation is usually stopped.

m30   Same as m02 in some N/C machines.

m50   ZERO CAM. No vertical control cycle is started. Movement is made only in X and Y, and spindle can be hand-operated for vertical travel. The tape stops. Push *cycle start* to continue.

m51–m59   CAM selections. To call for preset spindle travel. Rapid and feed vertical distances are set by the operator as part of the setup procedure. In some of the new machines the cam selection is specified in a separate "W" column. The cams are then numbered W1, W2, etc.

Note: Other numbers (for example, 90 to 99) are sometimes used for cam or vertical stop selection, and different makes of N/C machines use somewhat different coding. However, the above coding is a typical example and is used in the programs in this and the next chapter.

Z Axis Control

There is probably less standardization in the method of controlling the Z axis than in any other part of numerical control programming. None of the systems is especially difficult, and each is explained clearly in the instructions sent with each N/C machine. This book illustrates some typical systems.

A simple method of depth control which is used on some vertical-spindle machines is to manually set several pairs of cams. Each pair of cams controls:

1. The amount of downward motion at rapid advance, to the feed engagement point.
2. The depth of travel at a preset feed rate, called the feed depth.

Depending on the machine, these cams can be set to accuracies of from 0.04 to 0.12 mm. They will repeat on each part within somewhat closer limits. These N/C machines can be purchased with six or more pairs of cams.

A cam on an N/C machine, as shown in Fig. 6-5, may simply be a circular disk with a button on it. This button will operate a limit switch connected to the machine controls. The action of the spindle, due to opening or closing the limit switches, will vary according to the preparatory (g) signal given from the tape.

Figure 6-5 shows part of the action of a pair of cams. The two cams are actually close together. They are shown separated, so that the drawing will be easier to follow. If the signal is g81 (DRILL cycle), the cams and switches will operate as shown. The electric and electronic circuits are omitted, since this is a study in itself.

Points 1, 2, and 3 on the cams and at the spindle in Fig. 6-5 represent:

1. Position of all equipment at the start of the cycle. The start of the rapid traverse to the clearance line.
2. Positions at the end of the rapid advance. The beginning of the controlled feeding advance.
3. Position at the completion of the drilling. Start of the rapid return to position 1.

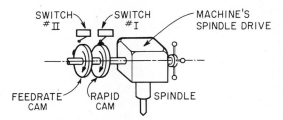

Schematic drawing of one pair of cams at position #2 of sketches below.

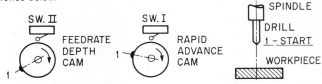

At 1. — Both switches open. Spindle starts down at rapid rate.

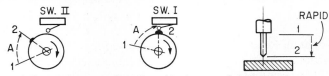

At 2. — Switch I closes. End of rapid advance. Spindle continues down at feedrate.

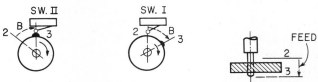

At 3. — Final depth of hole. Switch II closes. If drilling, spindle retracts to position 1 at rapid retract rate.

Fig. 6-5 Schematic drawing of one type of cam-controlled Z-axis (depth) movement.

At position 2 the **rapid** cam has rotated through angle A and contacted switch I, which cuts off the rapid traverse and starts the controlled downward feed. The feed cam has also rotated through angle A, but has not reached the switch. At position 3 the **feed** cam has rotated through angle B and contacted switch II. This stops all advance motion and starts the spindle back to its first position at rapid speed.

In operation the cams rotate with the shaft. They are first set by hand and secured at the correct position by set screws or special fastenings. Each pair of cams can be adjusted in a minute or two.

Some N/C machines use a rotating group of **micrometer** stops to control the Z motion; others use electronically adjusted stops. In each case the adjustment is quick, simple, and quite accurate. Of course, in true three-axis machines, these Z motions are controlled by the tape alone.

Preset Tooling

If cam m51 has been adjusted to drill the two 14-mm holes in the support base (Fig. 6-8) and the drill gets dull, a problem arises when the operator needs to replace it with another drill. Probably no two 14-mm drills that have been used in a shop are the same length. If a different-length drill is set into the spindle, the depth cam will have to be reset. There are, of course, several possible solutions to this problem:

1. If depth is not critical, the operator can simply measure, with a scale, how far the dull drill projects from the holder, and set the sharp drill to the same length.
2. There is sometimes a special light on the console of the N/C machine which shows when each cam is operating. This can help the operator to set the tools to length without actually measuring. Or an electric meter is nulled (set to zero) by dialing in the proper dimensions.
3. If the job is at all long or complicated, or calls for close tolerances on depths, most companies make up duplicate sets of tooling for these critical operations. This requires that some kind of record be available showing the setting length for each cutting tool. Some companies have established a system of cutter layouts or drawings for every cutting tool used on their N/C machines. These layouts are drawn up to describe the tool and its setting length, and each tool is given a number. On machines so equipped this is also the number used on the automatic tool changer.

Thus a 14-mm HSS standard jobber's length drill for this N/C machine might be set according to drawing 0532, as shown in Fig. 6-6. Then the tool crib or the machine operator simply assembles the drill in its holder according to the drawing **every time** this 14-mm standard drill is used.

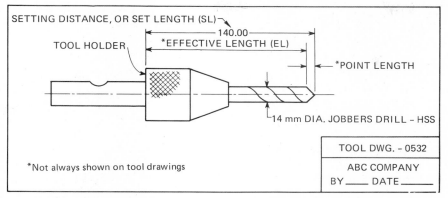

Fig. 6-6 Typical tool drawing, showing the setting distance. Type or make of toolholder may also be specified on some drawings.

4. For some short-run, relatively simple jobs, the N/C programmer (working with the operator or setup person) just makes a note on the programming sheet (or a separate tooling sheet). This might simply state "14-mm HSS drill set to 140.00 OAL" (overall length). The operator is furnished with equipment near his machine for quickly checking and setting these tools.
5. In milling operations, where depth is often held closely, proper setting of tool length becomes especially important. These tools are generally set within ± 0.02 mm, using special fixtures and gages.

    Machines are made with sensing devices, which, when the tool touches the part, automatically change the spindle feed from rapid to feed rate. On these machines, drills and taps can be left at random lengths.

Whatever system is used, it is important to **keep the machine cutting metal** as much of the time as possible. Thus some arrangement must be made so that dull cutting tools may be replaced quickly and still maintain the accuracy needed on the work.

Cutting Distance at Feed Rate

The feed rate during cutting is relatively slow, and, for efficient machining, this distance should be kept as short as possible. To accomplish this, rapid advance and retract of the spindle is normally used whenever the tool is not cutting. Several factors must be considered when deciding the length of the motion at feed rate. They are required clearance from the work, depth of cut, breakthrough allowance, and point allowance. The

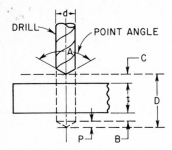

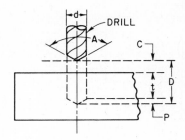

DRILLING THROUGH HOLES   DRILLING BLIND HOLES

A = point angle of drill
B = breakthrough allowance to en-
    sure fully drilled hole; decided
    by programmer
C = clearance from work to pre-
    vent a collision at rapid ad-
    vance rate; varies according
    to work and machine

d = drill (or countersink) diameter
D = distance drill travels at pro-
    grammed feed rate
P = point allowance; for standard
    118° drills, $P = 0.3 \times d$.
t = thickness of part, or hole
    depth, as shown on part
    drawing

From the sketches above, the formula for D is

$$D = t + C + B + P$$

Notes: **1.** B = zero for blind holes.
      2. C and B are selected by the programmer according to the N/C
          machine's capabilities, the roughness of the workpiece, and the
          programmer's judgment.
      3. P must be computed. For angles other than 118°, see Appendix
          A, item VII.

Fig. 6-7 Sketches and formula for computing depth of cut at feed rate, for
through and blind holes.

computations needed for setting the feed rate distance for drilled holes are
made as shown in Fig. 6-7. Some other considerations are as follows:

    1. A pair of cams may be used for several different tools if the tools
       have the same setting lengths and the same rapid advance and feed
       rate depths.
    2. Taps require special consideration because point angles and the
       number of partial threads vary according to the type of tap used.
       The tap drill depth should be shown on the drawing. However, it is
       wise for the part programmer to check this to be sure it is deep
       enough.
    3. If the design permits, it is sometimes more economical to drill
       "through" before tapping in order to avoid chip problems in blind
       holes.
    4. Sometimes the rough and finish milling cutters are preset to dif-
       ferent depths, so that they can use the same cam or depth stop set-
       ting.

Writing a Word Address Program

The **manuscript form** shown in Fig. 6-11 provides space for all the information needed on the N/C machine described in this chapter. The five columns between the double vertical lines represent information which will be punched into the tape. The next five columns supply information to the machine operator. These added columns tell the operator the tools, feeds, and speeds which must be set by hand for each section of the work.

A copy of this sheet is sent to the operator along with the tape, tools, and work. Sometimes under the "tool" column is a tool number, which refers the operator to a standard drawing describing the tool, the toolholder, and any special information, such as the length or special setting gages.

The WORD ADDRESS program for the drilling, tapping, and boring of

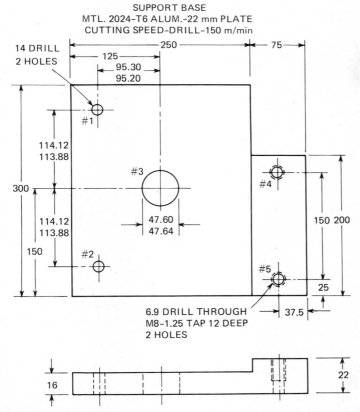

SUPPORT BASE
MTL. 2024–T6 ALUM.–22 mm PLATE
CUTTING SPEED–DRILL–150 m/min

Fig. 6-8 Original drawing of support base. Notice that some holes are dimensioned from the centerline.

the support base (Figs. 6-8 and 6-10) is shown in Fig. 6-11. Milling operations are discussed in Chap. 7.

The distance between the zero point and the setup point must be decided before starting to write the program. This usually requires consulting with the methods and/or production departments to plan the clamping or fixtures necessary for securing the workpiece to the machine table. Once this is decided, it is a good idea to make a sketch or a scaled drawing of the workpiece in position on the machine table. This layout should include the principal dimensions of the fixture and the necessary clamping in outline (not detailed). These details have been omitted from the drawing of Fig. 6-10 in order to concentrate on the programming.

Coordinate dimensions, from a convenient baseline, usually simplify the arithmetic needed for writing a part program. Notice that on the drawing (Fig. 6-10) the setup dimensions are given and that the part dimensions have been refigured so that they are now coordinate, or abso-

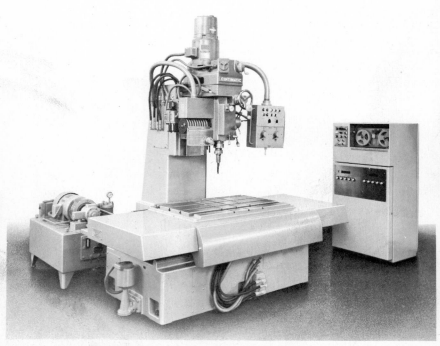

Fig. 6-9   A 2.2-kW [3-hp] two-axis vertical-spindle N/C drilling, milling, tapping, and boring machine. Fixed zero, FULL ZERO SHIFT, WORD ADDRESS programming. Cam-controlled depth (Z axis) called for by tape. Positions to ±0.02 mm [0.001 in.] in 610 mm [24 in.]. (*Cincinnati Milacron.*)

lute, dimensions from a convenient point. Of course, all points must finally be located by absolute dimensions from the fixed zero.

The designer of this part wanted to be certain that the two drilled holes were located 114.0 ± 0.12 mm [0.005 in.] from the center hole (Fig. 6-8). Thus the 114.12 to 113.88 dimensions are called for. However, the location of the center hole from the edges of the part are whole number dimensions (no decimals) of 150 and 125 mm. This ordinarily allows ± 0.40 mm [¹/₆₄ in.] which makes setup very easy.

Most N/C machines today will consistently locate the spindle or table within ± 0.02 mm or closer. However, actual working conditions require that allowance must usually be made for spindle runout, toolholder and tool runout, and "walking" of the tool as it touches the workpiece. Each of these variations is quite small, so that ±0.05 to ±0.12 mm tolerance can usually be maintained quite easily.

Thus the machine, using coordinate dimensions, will hold the required tolerance easily, and the use of the basic dimensions on our layout sketch makes program-writing a very simple job.

Figure 6-11 shows the complete program for the drill, tap, and bore work. The figures for rpm and feed are based on a cutting speed for drills of 150 m/min and a feed of 0.08 mm per revolution (mm/rev).

The way rpm and feed rates are figured is shown below.

$$\text{rpm} = \frac{300 \times \text{cutting speed}}{\text{drill diameter}}$$

$$\text{feed} = \text{rpm} \times \text{mm/rev} = \text{mm/min}$$

Holes 1 and 2   14-mm drill

$$\text{rpm} = \frac{300 \times 150}{14} = 3214 \text{ rpm, use 3200 rpm}$$

$$\text{feed} = 3200 \times 0.08 = 256 \text{ mm/min}$$

Holes 4 and 5   6.9-mm tap drill

$$\text{rpm} = \frac{300 \times 150}{6.9} = 6521 \text{ rpm, use 4000 rpm}$$

$$\text{feed} = 4000 \times 0.08 = 320 \text{ mm/min}$$

Holes 4 and 5   M8-1.25 tap

rpm = 352 rpm (See Appendix E)

feed = 352 × 1.25 = 440 mm/min

Note: Feed for tapping = pitch of screw thread.

Hole 3   44-mm drill

$$\text{rpm} = \frac{300 \times 150}{44} = 1023 \text{ rpm, use 1020 rpm}$$

$$\text{feed} = 1020 \times 0.08 = 81.6 \text{ mm/min, use 80 mm/min}$$

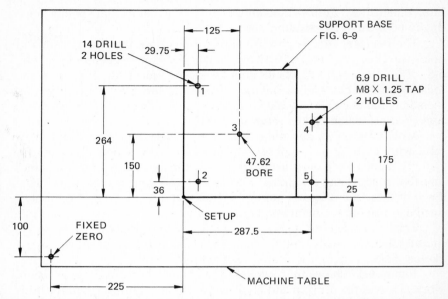

Fig. 6-10 Combined part and table-location drawing. Notice that the baseline dimensioning is now used, and setup dimensions are shown. All dimensions are in millimetres.

Hole 3   47.62-mm bore

$$rpm = \frac{300 \times 150}{47.62} = 945 \ rpm$$

Use 0.05 mm/rev feed for boring.
    feed = 945 × 0.05 = 47.25 mm/min, use 50 mm/min

Take special notice of the following specifications:
1. Any g, x, y, or m50 to m59 cam number command **remains in effect until a new command is given.** The M02 and m06 commands are canceled when a new block of tape begins.
2. The home, or fully retracted spindle position, must, in this machine, be high enough so that the 6.9-mm drill will not collide with the raised section when moving to hole 4 or 5.
3. All moves in *X* and *Y* are made before any vertical (*Z*) motion starts, except when milling with the g79 commands.

The Actual Program

Fig. 6-11, figured from dimensions in Fig. 6-10, is:

Hole 1   **n005 g81 X25475 Y36400 m51**
    005   The SEQUENCE or LINE NUMBER.
    g81   The DRILL cycle command.

$$X = 225.00 + 29.75 = 254.75 \text{ mm}$$
$$Y = 100.00 + 264.00 = 364.00 \text{ mm}$$

    m51   Depth cam number 1; hand-set by the operator.
Hole 2   **n010 Y13600 m06**
No change in g or X; so these can be omitted.

$$Y = 100.00 + 36.00 = 136.00 \text{ mm}$$

    m06   Tool change, after the drilling is completed.
Hole 5   **n015 g81 x51250 Y12500 m52**
(Hole 4 could have been drilled now. However, hole 5 is closer to the last hole drilled, and thus this move saves some travel time.)
    g81   Must be repeated after tool change.

$$X = 225.00 + 287.50 = 512.50 \text{ mm}$$
$$Y = 100.00 + 25.00 = 125.00 \text{ mm}$$

    m52   New cam, since depth settings are different for the new tool, and different work height.
Hole 4   **n020 Y27500 m06**
g81 and X dimensions do not change.

$$Y = 100.00 + 175.00 = 275.00 \text{ mm}$$

    m06   Tool change.
Hole 4   **n025 g84 X51250 Y27500 m53**
    g84   The tapping command.
No change in X or Y, but it is good practice to use a full block of information after every tool change, so that the program can be "picked up" here if a tool breaks or if there are other interruptions.
    m53   New cam set for proper tapping depth.
Hole 5   **n030 Y12500 m06**
No change in g84 or X coordinate.
Y = Same as in n015.
    m06   Tool change.

    The next sequence of five lines illustrates one way to machine the 47.60–47.64-mm diameter called for. Other equally good methods might be used.
Hole 3   **n035 g81 X35000 Y25000 m51**
    g81   DRILL command.

$$X = 225.00 + 125.00 = 350.00 \text{ mm}$$
$$Y = 100.00 + 150.00 = 250.00 \text{ mm}$$

m51   Cam used because drill is the same as used for holes 1 and 2.

Hole 3   **n040 g80 m06**

g80   Prevents any vertical (Z) movement.

m06   Must be on a new line because the machine will accept only one m command per block and the m51 was needed in line n035.

Notice that the m06 is written next to the g80. All blank spaces in this program could be omitted since the "words" give the address. However, it is much easier to check a program, if most of the items are kept in columns, by using the TAB key.

Hole 3   **n045 g81 m54**

g81   Repeated after tool change, no change in X or Y.

m54   The new cam for use with the larger drill.

Hole 3   **n050 g80 m06**

Same as n040

Hole 3   **n055 g85 m55**

g85   Bore command, feed rate both down and up.

No change in X or Y.

m55   Cam to control boring depth.

The next sequence is to move the spindle off the workpiece.

n060 g80 X19000 Y40000 m02

g80   Prevents any vertical (Z) movement.

$$X = 225.00 - 35.00 = 190.00 \text{ mm left of the work.}$$
$$Y = 100.00 + 300.00 = 400.00 \text{ mm back of the work.}$$

m02   Machine stop and tape rewind.

Note: Sequence n060 is optional. However, it is frequently wise to have the spindle out of the way while the operator is changing the workpiece. Occasionally, a similar move may be desirable before a tool change.

Spot Drilling

The material used in this problem is aluminum. Thus one of the special drill points can be used to prevent the drills "walking" out of true position. Depending on the allowable hole-location tolerance, if the material were steel or any of the harder or tougher metals, it might be necessary to spot-drill or center-drill before following the program just written. All the X and Y positions would be as shown, but of course only one g81 command, two cams (m56 and m57), and a final m06 would be needed. Thus it would require only five more lines, at the beginning of the manuscript, to add the spot drilling.

Extra Complete-Information Blocks

You may have wondered why, at SEQUENCE NUMBER n025, the complete information was programmed, even though the X and Y coordinates did not change. On a short tape such as this, it is not too important. However, with long tapes it is sometimes quite important.

One reason is that if the power to the N/C machine is turned off, the entire memory is wiped clean. This might happen at any time during the program. Or, during the trial run, it might be desirable to go back to re-machine a section of the workpiece; or during production, a boring tool may need to be reset, and the hole rebored.

In any case, the MCU requires full information in order to locate correctly and perform the desired operation. If this complete information is not given at fairly frequent intervals, it might be necessary to go back to the beginning of the program, or back an added five, ten, or a hundred blocks to pick up all the needed data.

Most N/C machines have a tape-rewind control on the operator's console. With this, and a SEQUENCE NUMBER READOUT, the operator can go back to a complete block (which he or she can locate on a copy of the program manuscript) and start up at this point.

The machine shown in Fig. 6-9 uses a letter H instead of the usual N

PART NO. FIG. 6-8 & 6-10	PART NAME SUPPORT BASE	MTL. ALUM. 2024-T6	PAGE 1	PAGES OF 1	DATE		BY			
OPERATION	SEQ. NO.	PREP FUNCT.	X POSITION	Y POSITION	MISC. FUNCT.	POS. NO.	SPEED RPM	FEED mm/min SPNDL	FEED mm/min TABLE	REMARKS
	SET-UP-FRONT L.H. CORNER AT X225.00, Y10000									
14 DR. -HOLE 1	n005	g81	x254 75	y364 00	m51	1	3200	256		14 HSS DRILL
-HOLE 2	n010			y136 00	m06	2				
6.9 DR. -HOLE 5	n015	g81	x512 50	y125 00	m52	5	4000	320		6.9 HSS DRILL
-HOLE 4	n020			y275 00	m06	4				
M8 TAP -HOLE 4	n025	g84	x512 50	y275 00	m53	4	352	440		M8-TAP HSS
-HOLE 5	n030			y125 00	m06	5				
14 DR. -HOLE 3	n035	g81	x350 00	y250 00	m51	3	3200	256		14 HSS DRILL
	n040	g80	m06			3				
44 DR. -HOLE 3	n045	g81			m54	3	1020	80		HSS STUB DR.
	n050	g80	m06			3				
BORE 47.62 DIA.	n055	g85			m55	3	945	50		BORING TOOL CARBIDE TIP
CLEAR WORK PIECE	n060	g80	x190 00	y400 00	m02	SPINDLE AT BACK L.H. CORNER				
		REMOVE WORK								
		CHANGE TOOL								
		SECURE NEXT PIECE								

Fig. 6-11   A WORD ADDRESS program for drilling, tapping, and boring the support base of Figs. 6-8 and 6-10. All dimensions are in millimetres.

Fig. 6-12 A worker wiring the back of the panel of the MCU shown in Fig. 6-9. (*Cincinnati Milacron.*)

to identify the sequence number of these blocks, and a *search* button on the MCU will rewind the tape to the nearest H block. Various methods are used to assist in locating these restart blocks. On a simple machine, the programmer may have five or six code delete punches inserted so the operator can visually locate the nearest complete block.

Of course, on N/C machines with a FLOATING ZERO, it is necessary to zero in the spindle again every time the power is cut off. After this is accomplished, the program can be restarted at any block which gives complete information.

Whatever method is used, the programming of the additional data will, in the long run, save considerable production time.

Behind the MCU Panel

As a matter of interest, Fig. 6-12 shows one method of wiring the back of the control panel of the machine control unit. The clock-like device tells the operator the direction in which to run the wires. The tubes contain wires precut to the proper lengths. Each tube is numbered. The readout, just below the "clock," tells the assembler the number of the tube in which the next wire is located. Thus, with these helps and a simple wiring diagram, the operator can quite speedily wire the "card cages" of the MCU.

Chapter 7

STRAIGHT-CUT MILLING

Most numerical control machines which drill, tap, etc., can also be used for a variety of milling operations. However, the point-to-point (NPC) machines discussed in this book can do only straight-cut, or picture-frame, milling. That is, they can be accurately directed in only a straight X or straight Y direction. Thus cuts or slots not parallel to X and Y cannot ordinarily be made on NPC machines, though a 45° angle can be closely approximated.

These machines are limited to straight cuts because, in order to make a milling cut along the hypotenuse of a 3-4-5 right triangle, the motors for X and Y axis movement must cause motion at speeds with exact ratios of 3:4. This requires variable-speed motor controls which are available only on the contouring (NCC) type of machine and on machines equipped with liner interpolation capabilities. Refer to Chap. 2 for further information on this.

Feed Rate Control

Making a milling cut requires control of the feed rate along the X and Y axes in addition to control of the Z axis feed, which was used in drilling. The X or Y feed is, on the simpler machines, hand-set using the *milling feed* dial. Thus, when the MILLING command (G code) is read from the tape, the feed rate which has been set on this dial is automatically used by the carriage or table movement. This feed rate is usually in millimetres per minute. Many of the more sophisticated N/C machine controls will accept feed rates which have been coded into the tape using an "f" code.

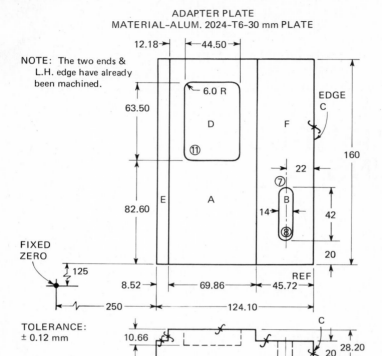

Fig. 7-1 Part drawing, showing location from fixed zero.

Types of Milling Cuts

Facing a flat surface, or the tops of a series of ribs in a part, may be done with an end mill, face mill, shell mill, or fly cutter (Fig. 7-1 at *A*). This facing operation frequently requires several passes of the cutter, since there is a limit to the diameter of cutter which can be used on a given machine. This limit will depend on the material being cut, the size and rigidity of the machine spindle, the horsepower of the machine, and the accuracy and finish desired.

When making a facing cut it is usually good practice to have the cutter go completely off the workpiece at the end of each pass across the work.

Grooves, keyways, and slots (Fig. 7-1 at *B*) may be made with a cutter which has the exact diameter desired. That is, a 6-mm end mill will often be used to cut a 6-mm-wide keyway. If the width of the cut is not the same as a standard cutter diameter, or if the width must be held to close tolerances, two cuts may be made. The second pass will be offset from the first by the difference between the cutter diameter and the width desired. Both the Z- and the X- or Y-axis feed controls can be used in this type of cut.

Edge cuts (Fig. 7-1 at C) on the outside edges can be finished by using the side-cutting capabilities of end mills or shell mills. Or a notch may be cut in some or all of the edges of a part, as shown in Fig. 7-1 at E. This notch can be a 90° cut or rounded, etc., according to the style of milling cutter used. Feed methods will be the same as in facing, and it is good practice to program the cutter to start and stop while it is off the workpiece at each end.

Pocket milling and window cutting (Fig. 7-1 at D) are cuts used to mill out a pocket in part of the surface of a workpiece. These may be square or rectangular, and may be made to any depth required. If a window completely through the part is required, it is sometimes necessary to make several cuts at different depths until the last cut can go completely through.

Pocket milling can be programmed in a number of different ways, as shown in Figs. 7-2 to 7-4. A limiting factor is that the final cuts must be made with a cutter which has the same radius as the corner radius shown on the drawing. For example, in the drawing of Fig. 7-2, the radius shown is 10 mm, so the final cutter must be a 20-mm-dia. end mill. Three possible methods of pocket milling are as follows:

1. The use of **rough and finish cutters** of different diameters can be a very efficient method, especially if the corner radius of the pocket is small. In the example shown in Fig. 7-2, a 32-mm 2-flute roughing cutter was used, leaving 0.8 mm stock along the sides to be cleaned up by the 20-mm finishing cutter. The roughing cuts overlap each other to avoid leaving a thin ridge on the work.

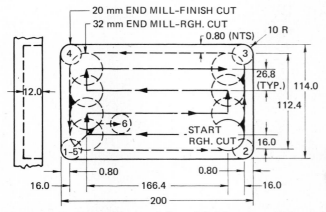

Fig. 7-2 Pocket milling, rough and finish cuts, with two sizes of cutters.

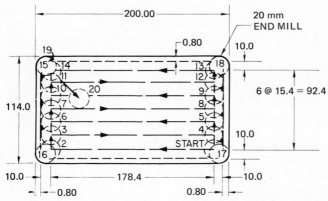

Fig. 7-3   Pocket milling, using one cutter, with final pass cutting to finished size.

Notice that the roughing cutter leaves "scallops" in several places. This excess material must be small enough so that the finishing cutter will remove it.

The roughing cuts may start at any point inside the finish line, and proceed in either the X or Y direction. It is easily seen that fewer tape commands will be needed if the first cut goes along the longest dimension of the pocket.

It is sometimes important to consider carefully the direction in which the cutter travels. Most N/C machines use ball lead screws, or have backlash take-up of some kind, so climb (down) milling is no problem. On many materials, especially on aluminum, climb milling gives a much better finish than conventional (up) milling, and it is less likely to dig in at the corners and make a gouge. Good machine-shop practices should be followed; so materials such as cast iron, which has a tough scale, would probably be cut using conventional (up) milling.

To make a climb cut in a pocket, using a standard right-hand milling cutter, the programmer must have it go around the pocket in a counterclockwise (CCW) direction. In Fig. 7-2 the successive positions of the 20-mm-dia. finishing cutter are numbered 1 to 5 in counterclockwise order. Position 6 is programmed to bring the cutter away from the wall of the pocket, into the previously machined area, before raising it out of the pocket. This helps to avoid a dwell mark on the finished surface of the wall of the pocket, and the move at an angle is permissible because no machining is being done.

This particular example (Fig. 7-2) requires locating 8 points (4 passes) for roughing and 6 points for finishing, plus the time required to change the cutter. Total distance traveled is about 1300 mm.

2. **Using one cutter** for the entire operation might be economical for this particular pocket-milling job. Figure 7-3 (same outside dimensions as Fig. 7-2) shows how this is done. Use of the 20-mm-dia. cutter requires more passes to clean out the center. The finishing cut is the same as was used in Fig. 7-2.

   Thus this method requires locating 14 points (7 passes) for roughing and 6 points for finishing, with no tool change required.

   Notice that if the corner radius specified on the drawing were quite small (3- or 6-mm radius, for example), this method would require too many passes back and forth, and would be impractical, because of the time required.

   In both the above examples, the dimensioning of the points required to locate the cutter centers is quite simple. There are only two X dimensions each for rough and finish cuts, and the Y dimension of the roughing cut usually varies by a fixed distance (15.4 mm in the example shown in Fig. 7-3).

3. **Starting the cut near the center** of the pocket and working outward (as shown in Fig. 7-4) is a method which may not require an extra finishing pass. This method is especially effective if the pocket is large and the bottom wall, or web, is thin. Under these conditions the center of the pocket will tend to bend because of the pressure of the cut. When the center cut is taken first, while the rest of the pocket is still full thickness, bending is very unlikely.

   In Fig. 7-4 the last pass around removes 10 mm of stock in order to keep the number of passes at a minimum. This would not be advisable if a close tolerance were required. Notice the coun-

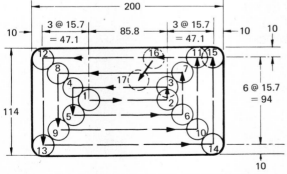

20 mm DIA. END MILL USED FOR ENTIRE POCKET

Fig. 7-4 Pocket milling, one cutter, starting at the center of the pocket.

terclockwise direction of the cutter travel. This results in climb milling cuts.

Location 16 needs only to be far enough beyond location 15 to remove the small scallop left by the cutter at position 11. The move to location 17 is made as in previous examples, to move the cutter into the clear before raising it, and it may be any direction away from the surface, and a relatively small distance.

This method requires locating 17 cutter center points, three fewer than required in Fig. 7-3. However, the computation of these center points will require the programmer to do more work, which is a disadvantage of this system. If a finish cut of 0.8 mm were specified in Fig. 7-4, one more cut all around (4 more points) would be needed.

In the last two examples, the 20-mm cutter travels a total of approximately 1800 mm in Fig. 7-3 and approximately 1400 mm in Fig. 7-4 (or about 1950 mm if the extra pass is needed).

End mills with a radius at the bottom require careful planning. As shown in Fig. 7-5, a 25-mm-dia. end mill with a 6-mm radius is removing only 13 mm of stock on the flat. Thus, if this shape cutter is used in multiple passes, it can only "step over" less than 13 mm each pass. This situation is common when the drawing calls for a radius at the bottom corners of a pocket.

If the cutter is advanced slightly less than one-half its diameter (for example, 0.4 of the diameter), there will not be any scallop, and the method shown in Fig. 7-2 will do the whole job. This is usually not desirable since it requires more cuts and takes too long.

The N/C programmer must decide which method will be the best in each situation, taking into consideration the accuracy needed, size of the pocket, thickness of the work, finish required, corner radius, and the time needed for machining by each method.

Spot-face and counterbore cuts, using end mills or counterbores, can be made by calling for the regular DRILL command (g81)

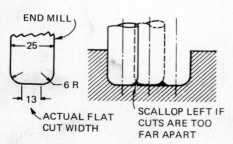

END MILL
25
6 R
13
ACTUAL FLAT CUT WIDTH
SCALLOP LEFT IF CUTS ARE TOO FAR APART

Fig. 7-5 Effect of milling-cutter radius on area milled out.

or with MILLING commands. The regular Z-axis feed rates are used. Some N/C machines are equipped so that a dwell may be programmed at the bottom of these cuts, in order to improve the finish.

Preliminary Information Needed for Milling Program

The machine to be used for this sample program is the same one used in Chap. 6. The specifications are as follows:
1. Single-vertical-spindle N/C machine.
2. Drill, bore, mill, and tap have canned cycles, called for by the proper G commands.
3. Depth is controlled by hand-set cams. Nine pairs of cams are available.
4. Speeds and feeds are set by hand. Vertical Z axis and horizontal (milling) feeds are on separate controls. Maximum 4000 rpm and 760 mm/min feed.
5. Fixed zero at front left-hand corner of machine table.
6. WORD ADDRESS VARIABLE BLOCK tape format.
7. See Table 7-1 for a list of the G and M code numbers, which are repeated for your convenience. Refer to Chap. 6 for full descriptions.

Referring to the drawing of Fig. 7-1, it is necessary to decide the order of operations, the cutting tools to be used, and how the work will be fastened to the machine table. There are several ways each of these items can be planned, and no one is necessarily the "right" one. However, the N/C programmer, in cooperation with others, must decide on one method, one specific set of cutting tools, and one fixture.

Table 7-1 SUMMARY OF PREPARATORY (G) CODES AND MISCELLANEOUS (M) CODES FOR N/C MACHINE USED IN CHAPS. 6 AND 7

G code	Meaning	M code	Meaning
g78	Mill cycle stop	m00	Program stop
g79	Mill cycle	m02	End of program
g80	Cancel cycle	m06	Tool change
g81	Drill cycle	m50	Zero cam
g84	Tap cycle	m51	
g85	Bore cycle	to	
		m59	Depth cam selections

For this workpiece (Fig. 7-1), it was decided to finish and square up three edges of the part before placing it on the numerical control machine. This can be done on standard equipment. Finishing the right-hand edge to the 124.10 dimension might also have been done on other machines. However, the N/C machine can easily finish this edge to tolerance, and it is a good illustration of how this type of cut is programmed.

Tooling Used (for Workpiece in Fig. 7-1)

A simple vise-type **fixture** can be designed to hold this part. The sides of the vise would be 13 to 16 mm high; so they will be below the shelf (F) and notch (E) cuts.

Cutting Tools

> 12-mm 2-flute HSS end mill (finish pocket D)
> 25-mm 2-flute HSS end mill (edge C, and roughout pocket D)
> 32-mm 4-flute HSS end mill or a shell end mill (face surfaces A and F and cut notch E)
> 14-mm 2-flute HSS end mill (slot B)

Order of Operations

1. Face surface F to $8.52 + 69.86 = 78.38$ shoulder dimension and 20-mm-high surface
2. Face surface A
3. Cut notch E
4. Finish edge C
5. Rough out pocket D
6. Finish pocket D
7. Cut 14 by 42 slot, 20 mm deep plus breakthrough

Note: This order of operations is not the only one, or even the best one. It is chosen to illustrate different items used in programming.

Since no close tolerances are called for, no finish cuts are needed on this part. However, when required to hold close tolerances, it is frequently necessary to make rough and finish cuts, leaving only 0.2 to 0.8 mm to be removed by the sizing pass.

Milling-Cutter Location

It is important to remember that the N/C machine will locate the **center of the milling cutter at the coordinates given** on the tape. The machine does not "know" where the edge of the cutter is working; so the programmer must do some careful arithmetic in planning the program. It frequently saves time, and certainly increases accuracy, if the programmer makes

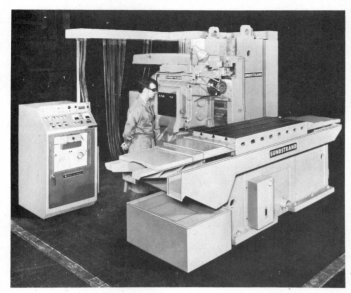

Fig. 7-6 A heavy-duty 15-kW [20-hp] N/C milling center. Also drills, taps, and bores. Table moves left and right (*X* axis); head moves up and down (*Y* axis); head moves in and out (*Z* axis). Available with circular interpolation. WORD ADDRESS programming, fixed zero. (*Sundstrand Machine Tool Division, Sundstrand Corp.*)

sketches similar to those shown in this chapter. All sketches should be saved for future reference and possible "debugging" of the tape. Any dimensions or notations placed on the print of the part drawing show up much more clearly if they are added in colored pencil or colored ink.

The Program for Milling (Figs. 7-1 and 7-7, Adapter Plate)

Figure 7-7 shows the part as it would look on the machine table. All dimensions have been refigured so that they are measured from two baselines. Using this new layout, the absolute (or coordinate) dimensions from the fixed zero can now be easily computed.

Face-milling section *F* is to be done with a 32-mm-dia. end mill. Two passes will be needed. The second pass will cut to the 78.38 shoulder dimension and, for good finish on the vertical edge, it will be climb milling. To climb-mill, this cut must start at the top. Good machine-shop practice requires that successive passes overlap each other, and the first pass must place the edge of the cutter outside of the right-hand edge.

With this information we can compute the locations of the **center** of the cutter at each position. Let us examine the dimensions shown in Fig. 7-8.

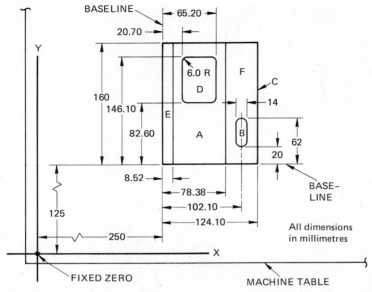

Fig. 7-7   Part drawing of Fig. 7-1, redrawn with baseline dimensioning, shown with setup dimensions to fixed zero.

The *Y* dimensions are as follows:

1.26   at the top and bottom of the cut is an arbitrarily assigned clearance dimension. It will vary according to the roughness of the edge of the part and the N/C programmer's judgment. It should be small, as it represents noncutting time.

16.0   at the bottom of the drawing is the radius of the 32-mm end mill. This is included to locate the center of the cutter.

17.26 = 1.26 + 16   This is the *Y* distance from the top or bottom of the workpiece to the center of the cutter. This figure can be used any time the 32-mm end mill is used for face milling on this part.

177.26   is the *Y* dimension from the workpiece baseline to the cutter at its top position. The 160 dimension plus the 17.26 just figured gives 177.26 mm. Baseline dimensions are used because they can easily be changed to program figures simply by adding the setup dimensions.

−17.26   is the *Y* dimension to the bottom position of the cutter. It is negative because it is below the baseline.

The last two, and future *X* and *Y* dimensions, have been "boxed" on the drawings so they will show clearly, as these are the dimensions to be used in figuring the final *X* and *Y* values for the program.

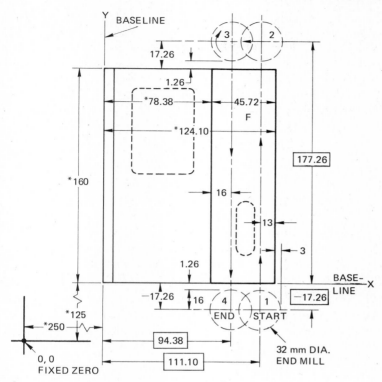

*Dimensions are from Fig. 7–7

Boxed dimensions are computed ℄ distances from
the baselines in Fig. 7–7. Add set–up dimensions
to get X and Y figures for the program.

Fig. 7-8 Dimensioned sketch for face milling surface *F*.

The *X* dimensions are as follows:

3.0 on the right side of the part is an arbitrary dimension which makes certain that the edge of the cutter projects to the right beyond the rough edge, which is to be finished later. Once again, this might be from 1 to 14 mm according to the work and the judgment of the programmer.

13.0 is the 16.0-mm radius of the cutter minus the 3.0 above.

$$16.0 - 3.0 = 13.0 \text{ mm}$$

This is the cutter centerline from the right-hand edge.

111.10 is the right-hand-edge dimension from Fig. 7-7 minus the above cutter centerline distance.

$$124.1 - 13.0 = 111.1 \text{ mm}$$

This is the baseline dimension for the start of the cut. Notice that this first cut removes a 29-mm-wide strip of material. Therefore the second cut, which has to hold to the 78.38 dimension, removes only $45.72 - 29.0 = 16.72$ mm of stock. This is a lighter cut, which will more easily hold to dimension.

16   from the left edge of this surface is the cutter radius. This location cannot be varied, since it is based on finishing to the 78.38 dimension.

94.38   is the X dimension of the cutter centerline for the second pass.

$$78.38 + 16 = 94.38 \text{ mm}$$

The X axis centerline dimensions of 111.10 and 94.38 are used at both the bottom and top of the facing passes. Notice that once the clearance dimensions of 0.8 and 3.0 have been decided, the rest of the figuring is simple. These figures should be written on the drawing, or a sketch of a part of the drawing, as shown in Fig. 7-8.

With the cutter centerline's X and Y dimensions clearly marked on Fig. 7-8, it is now quite easy to write the program for face-milling surface F. Refer to the m and g numbers shown in abbreviated form in Table 7-1. The program is shown in Fig. 7-9.

Note: Today, with automatic quill clamping frequently used, the g78 milling code is not often needed. However, there are a few hundred older N/C machines still giving good service, so the use of the g78 code is illustrated in some cuts in this chapter.

OPERATION	SEQ. NO.	PREP FUNCT.	X POSITION	Y POSITION	MISC. FUNCT.	POS. NO.	REMARKS
NUMERICAL CONTROL PROGRAM WORD ADDRESS – VARIABLE BLOCK FORMAT							
SET UP ON FRONT L.H. CORNER OF							
PART, AT X=250.00, Y=125.00 mm							
FACE MILL "F"	n001	g78	x36110	y10774	m51	1	CLAMP QUILL
(SEE FIG. 7-8)	n002	g79		y30226		2	(32 mm CUTTER)
	n003		x34438			3	
	n004			y10774		4	
	n005	g78				4	UNCLAMP QUILL
	n006	g81				4	SPINDLE UP

Fig. 7-9   WORD ADDRESS program for face milling surface F.

Explanation of Fig. 7-9

n001

g78   The table positions to the programmed X and Y point at rapid traverse. It feeds down to full depth, stops, and the clamp-quill light is turned on. The operator clamps the quill to keep the depth accurate, since this is a fairly deep cut. Push *cycle start* to continue.

$$X = 250 \text{ (set-up dim.) } + 111.10 = 361.10 \text{ mm}$$
$$Y = 125 \text{ (set-up dim.) } + (-17.26) = 107.74 \text{ mm}$$

m51   Will control all depths until a new cam number is programmed.

n002

g79   Starts the milling operation at milling feed rate. X remains unchanged as the table moves to $Y = 125 + 177.26 = 302.26$ mm.

n003   The only change is to move left to $X = 250 + 94.38 = 344.38$ mm. All other commands remain the same.

n004 X remains the same, and Y is the same as in line n001.

n005

g78   Stops the table and the tape, turns on the clamp-quill light. Operator unclamps the quill and presses the *cycle start* button to continue the work.

n006

g81   Causes the spindle to rise to home position at rapid rate.

Note: This program requires the use of the tab symbol if it is to be typed out in columns as shown. However, using the WORD ADDRESS type of programming, this is not necessary. The program could be written, or punched out, as:

n001g78X36110Y10774m51
n002g79Y30226
n003X34438
n004Y10774
n005g78
n006g81

Regardless of whether or not the tab is used when punching the tape, it is wise to write the program manuscript in an orderly arrangement, usually on a printed form such as used here or as shown in Fig. 7-22.

Face-milling surface A (Fig. 7-10) could be started at either the left or right side of the area. Since the 32-mm cutter is still being used and is now to the right of the area, it is logical to do the milling in the order shown in Fig. 7-10.

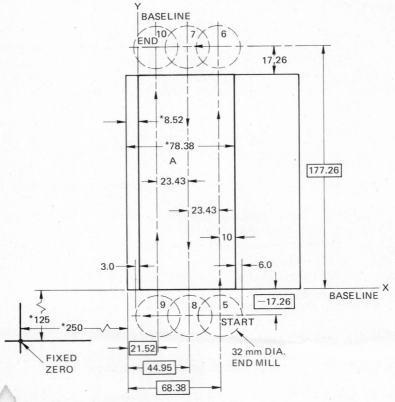

Fig. 7-10 Dimensioned sketch for face milling surface *A*.

*Dimensions are from Fig. 7-7

Boxed dimensions are from part baseline. Add set-up dimensions.

Y dimensions on Fig. 7-10 are the same as for milling the shelf *F*.

X dimensions are as follows: It is obvious that it will take three passes of a 32-mm cutter to finish a 70-mm-wide surface. The programmer must decide how much the cutter should extend beyond the edges of the surface, and how much each pass should overlap the previous one. The dimensions shown (6.0 beyond the right-hand edge and 3.0 beyond the left-hand edge) are no better than a number of others.

When these dimensions have been decided, the cutter centerlines may be assigned so that the overlap is approximately the same, or so that a little mathematics will make it the same for all passes, though this is not necessary.

In this example,

Total cutter width, three passes $\qquad$ $3 \times 32 = 96.00$ mm
Total width to be covered, surface width plus clearance
$78.38 - 8.52 + 6.0 + 3.0$ $\qquad$ $= 78.86$ mm
Difference available for overlap $\qquad\qquad\qquad$ 17.14 mm

Since three passes will have two overlapping edges, each overlap is:

$17.14 \div 2 = 8.57$ mm

Thus center-to-center distance of cutters will be the diameter of the cutter minus overlap:

$32.00 - 8.57 = 23.43$ mm

This is shown in Fig. 7-10.

> *Note:* As this part is cut from solid plate, it might be better to face all the way to the left edge. However, this cut illustrates a typical facing sequence.

The other dimensions shown in Fig. 7-10 are:

68.38 Cutter centerline at the start. This is the cutter radius distance from the 6-mm clearance. 78.38 (to right side) + 6.0 (clearance) − 16 (radius) = 68.38 mm + 250 set-up = 318.38 X dimension (or use 16.0 radius − 6.0 cl. = 10 mm, and 78.38 − 10 = 68.38 mm).

44.95 The previous centerline dimension minus the c-c dimension computed above, 68.38 − 23.43 = 44.95 mm + 250 setup = 294.95 X dimension.

21.52 Computed as above: 44.94 − 23.43 = 21.52 mm + 250 setup = 271.52 X dimension.

These last three figures are the programmed dimensions shown in Fig. 7-11. This cut is only 32-mm stock − 28.20 finish size = 3.80 mm (0.150 in.) deep, so it is unnecessary to clamp the quill. The program is shown in Fig. 7-11.

OPERATION	SEQ. NO.	PREP FUNCT.	X POSITION	Y POSITION	MISC. FUNCT.	POS. NO.	REMARKS
FACE MILL "A"	n007	g79	x31838	y10774	m52	5	32 mm CUTTER
(SEE FIG. 7-10)	n008			y30226		6	
	n009		x29495			7	
	n010			y10774		8	
	n011		x27152			9	
	n012			y30226		10	
	n013	g81				10	RAISE SPINDLE

Fig. 7-11 Program for face milling surface *A*.

Explanation of Fig. 7-11

n007

g79   Will move the table **at feed rate** to the new X and Y dimensions, rapid and feed down, and immediately call for the next tape command, which will, in this case, start the facing cut.

If the new X and Y positions were quite a distance away from the last position, a g80 command would move the table at rapid traverse to the new point. The next command would then use a g79 with no X or Y dimensions to lower the cutter, and a third block of information would start the cutter moving to the new X or Y dimension.

m52   Will select a new cam which has been preset (by the operator) at the proper depth for this operation.

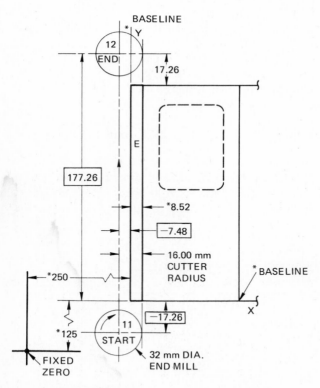

*Dimensions are from Fig. 7-7

Boxed dimensions are ℄ distances from baselines in Fig. 7-7. Add set-up dimensions.

Fig. 7-12   Dimensioned sketch for milling notch E.

n008–n012

> The successive X and Y dimensions are figured in the same manner as before. Notice that the Y dimensions simply alternate. If the cut is made from left to right, there will be only two X dimensions, used alternately.

n013 g81 As before, rapidly raises the spindle to the top position.

Cutting notch E (Fig. 7-12) could be done with a smaller cutter. However, it is a fairly deep cut and the 32-mm cutter is in the spindle, so it might as well be used. To get a good finish on the vertical edge, climb milling should be used. This requires starting the cut at the front (operator's side) of the part, as shown in Fig. 7-12.

Y dimensions are the same as in the previous facing cuts.

The X dimension is easily figured, since the edge of the cutter must be on the 8.52 surface shown in Fig. 7-8; so the cutter radius is subtracted. Thus $X = -16.0 + 8.52 = -7.48$. The minus sign indicates that the cutter center is to the *left* of the base line (which has actually been a temporary Y axis). The N/C program is shown in Fig. 7-13.

Explanation of Fig. 7-13

n014

> g78 Moves the table at rapid traverse to the new (X, Y) position, and positions to proper depth, off the work; clamp-quill light turns on.

> m51 Uses the same cam used for surface F because depth and cutter are the same in both cases.

> > X and Y coordinates are figured as before, by adding the setup dimensions to the baseline dimensions just listed.

$$X = 250.00 - 7.48 = 242.52 \text{ mm}$$

n015

> g79 Starts the milling cycle and moves the table to position 12.

OPERATION	SEQ. NO.	PREP FUNCT.	X POSITION	Y POSITION	MISC. FUNCT.	POS. NO.	REMARKS
MILL NOTCH "E"	n014	g78	x24252	y10774	m51	11	CLAMP QUILL
(SEE FIG. 7-12)	n015	g79		y30226		12	
	n016	g78				12	UNCLAMP QUILL
	n017	g81			m06	12	TOOL CHANGE

Fig. 7-13 Program for milling notch E.

n016

  g78   Stops the tape reader and turns on the clamp-quill light. Operator unclamps quill, pushes *cycle start* button.

n017

  g81   Raises the spindle to home position.

  m06   TOOL CHANGE. Stops the machine, and lights the tool-change light.

All the machining operations just programmed would be done by this numerical control machine without any stops except to allow the operator to lock and unlock the quill. Once the operator had set the feeds and speeds and pushed the *cycle start* button, all motions of the table and the spindle would be controlled by the tape.

Now the machine has stopped, and the operator will change to a 25-mm end mill in the spindle.

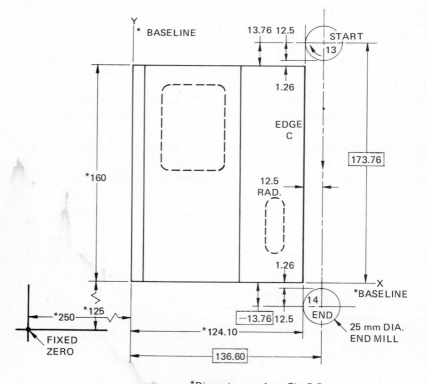

*Dimensions are from Fig. 7-7

Boxed dimensions are ℄ distances from baselines in Fig. 7-7; Add set-up dimensions.

Fig. 7-14   Dimensioned sketch for milling edge *C*.

Milling edge *C* (Fig. 7-14) will be done with a 25-mm-dia. 2-flute end mill. This could have been done with the 32-mm cutter, but the 25-mm is used in order to introduce the tool change and to get more practice in the computation of center distances. The computed dimensions in Fig. 7-14 are as follows:

1.26 Clearance, front and back. Arbitrarily assigned as before.

12.5 Cutter radius. Added to both *X* and *Y* dimensions to get cutter center dimension.

−13.76 The lower *Y* dimension is below the baseline; therefore it has a minus sign, $-12.5 + (-1.26) = -13.76$ mm.

173.76 The upper *Y* dimension is made up of 160 from Fig. 7-8 plus the clearance and the cutter radius.

$$160 + 1.26 + 12.5 = 173.76 \text{ mm}$$

136.60 The top and bottom *X* dimensions are the same, and this is fixed, since the cutter edge must touch the 124.10 dimension (from Fig. 7-7).

$$124.10 + 12.5 = 136.60 \text{ mm}$$

Once again, climb milling is required for good finish on aluminum. This is a light cut, and a depth dimension is not being held; so clamping the quill is not necessary. The milling program is shown in Fig. 7-15.

Explanation of Fig. 7-15

n018

g80   Used to bring the table to the programmed *X* and *Y* positions at rapid traverse.

m53   Selects a new cam, set for the deeper cut.

n019

g79   The spindle will rapid and feed to depth, and then read the next command.

n020

The cutter will move at the set milling feed, to the bottom *Y* dimension.

OPERATION	SEQ. NO.	PREP FUNCT.	X POSITION	Y POSITION	MISC. FUNCT.	POS. NO.	REMARKS
MILL EDGE "C"	n018	g80	x38660	y11124	m53	13	POS. @ RAPID
(SEE FIG. 7-14)	n019	g79				13	QUILL DOWN
	n020			y29876		14	
	n021	g81				14	QUILL UP

Fig. 7-15 Program for milling edge *C*.

n021
   g81   The spindle will rapid up to home position.

   Cutting pocket *D* (Figs. 7-16 and 7-17) requires two steps, roughing and finishing. The roughing should be done with large cutters to save time. In this case we already have a 25-mm-dia. end mill in the machine, and it is well suited for this job.
   The rough cutter is programmed so that it will leave stock all around the pocket, to be removed by the finishing cutter. The amount of stock to be left is decided by the N/C programmer, according to good machining practice. On hard-to-cut materials, the finish cut might be only 0.20 to 0.40 mm. On our aluminum part 0.80 mm is left for the finish cut.
   The diameter of the cutter for the finish cut must be twice the corner

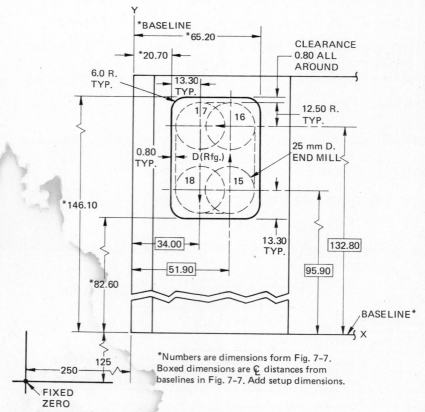

Fig. 7-16 Dimensioned sketch for rough milling pocket *D*.

radius shown on the drawing. For this part, the 6 mm radius requires a 12-mm-dia. end mill.

> *Note:* If the depth of the cut is greater than the diameter of the cutter, special care must be taken to maintain the drawing dimensions and to avoid breaking the cutter. Precautions would include slower feed, less stock left for finishing, and, in extreme cases, milling only the corners with the small end mill and cutting the straight sides to dimension with a larger-diameter cutter.
>
> On large work, where the radii called for may be 50 mm and over, a different problem exists. It is, for example, seldom possible to use a 150-mm-dia. milling cutter for finishing to a 75-mm radius. In this case a smaller cutter would be used, and programmed (on an NCC machine) to "contour"-cut an arc with a 75-mm radius. These large pieces are frequently programmed in APT III or a similar computer language.

The dimensions on Fig. 7-16 for the roughing cut are as follows:

0.80   Left for finishing cut all around the pocket.
12.5   Cutter radius.

The centerline dimensions for the rough machining are calculated by adding or subtracting $12.5 + 0.8 = 13.30$ mm from the appropriate baseline dimension.

The X dimensions to cutter centerlines are (to the baselines):

$65.20 - 13.30 = 51.90$ mm on the right
$20.70 + 13.30 = 34.00$ mm on the left

The Y dimensions to cutter centerlines (to the baseline):

$146.10 - 13.30 = 132.80$ mm at the top
$82.60 + 13.30 = 95.90$ mm at the bottom

This cut is started at the front right-hand corner because this is closest to the last position of the cutter. Notice that in the finish cut in Fig.

OPERATION	SEQ. NO.	PREP FUNCT.	X POSITION	Y POSITION	MISC. FUNCT.	POS. NO.	REMARKS
ROUGH CUT POCKET "D"	n022	g80	x30190	y22090	m54	15	RAPID TO POS.
	n023	g79				15	LOWER QUILL
(SEE FIG. 7-16)	n024			y25780		16	
	n025		x28400			17	
	n026			y22090		18	
	n027	g81			m06	18	RAISE QUILL TOOL CHANGE

Fig. 7-17   Program for rough milling pocket *D*.

7-16 there is quite a lot of extra material to be removed at the corners, and there is a "scallop" at the front which must be milled out.

> *Note:* If the drawing specified a radius at the bottom of the pocket, the 12-mm-dia. cutter with a radius might not clean out all this material; so an extra pass would be required. A carefully scaled drawing of this area, or some mathematical analysis, would be needed in order to know exactly what to do.

The program for the roughing operation is shown in Fig. 7-17.

Explanation of Fig. 7-17

n022

> g80   Moves the table at rapid traverse to the new (X, Y) position. Spindle stays in retracted position. Reads next block.

> m54   Selects a new preset depth cam.

n023

> g79   Lowers the quill at rapid and feed rate. Reads next block.

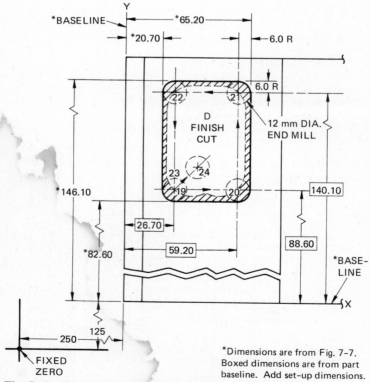

Dimensions are from Fig. 7-7. Boxed dimensions are from part baseline. Add set-up dimensions.

Fig. 7-18   Dimensioned sketch for finish milling pocket *D*.

n024–n026

Specify cutter positions taken from Fig. 7-16, plus setup dimensions.

n027

g81   Raises the spindle at rapid rate.

m06   Tool change. Stops machine and tape and turns on the tool change light.

Finish milling the pocket (Fig. 7-18). Climb milling is used; so the cutter must go around the pocket in a CCW direction. The program for this operation is shown in Fig. 7-19.

The edge of the cutter must touch the outside pocket dimensions; so the X and Y centerline dimensions are figured from the baseline dimensions in Fig. 7-7 by adding or subtracting the cutter radius, which is 6 mm.

The X dimensions to cutter centerlines are (Fig. 7-18):

$$65.20 - 6.00 = 59.20 \text{ on the right}$$
$$20.70 + 6.00 = 26.70 \text{ on the left}$$

The Y dimensions to cutter centerlines are

$$146.10 - 6.00 = 140.10 \text{ at the back}$$
$$82.60 + 6.00 = 88.60 \text{ at the front}$$

Explanation of Fig. 7-19

n028

g79   Moves cutter at feed rate to lower left-hand corner (it only moves 0.80 mm), lowers quill into work.

m55   Selects new depth cam because tool length is different.

n029–n032   Specify cutter centerline dimensions from Fig. 7-18, plus setup dimensions.

OPERATION	SEQ. NO.	PREP FUNCT.	X POSITION	Y POSITION	MISC. FUNCT.	POS. NO.	REMARKS
FINISH MILL POCKET "D"	n028	g79	x27670	y21360	m55	19	POS. @ FEEDRATE
(SEE FIG. 7-18)	n029		x30920			20	
	n030			y26510		21	
	n031		x27670			22	
	n032			y21360		23	
	n033		x30000	y24000		24	CUTTER IN CLEAR
	n034	g81			m06	24	RAISE QUILL TOOL CHANGE

Fig. 7-19   Program for finish milling pocket D.

n033   Specifies X and Y dimensions which bring the cutter into the previously machined area. Both X and Y dimensions are changed at the same time, which is not usually done. However, no cutting is being done, so the exact cutter path is not important.

Note: Moving the cutter into the clear before raising it out of the cut avoids enlarging the corner (sometimes called "dogboning"), which is caused by relieving the pressure on the cutter.

n034   g81 and n06   As in rough milling.

Cutting the slot B (Fig. 7-20) will require several passes at different depths, since a 14-mm-dia. end mill will not satisfactorily make a cut 20 mm deep. Direction of travel is not important since the cutting is being done by both sides of the cutter. But the finish may not be very good on one side.

Three cuts, each 8.0 mm deep will be used. The third pass will project 4.0 mm below the bottom of the part, making certain that the slot is straight through. Total depth is 24 mm.

Centerline dimensions for this slot are simple to calculate. The X dimension is given in Figs. 7-7 and 7-20 as 102.10 mm. The Y dimensions are:

Top $Y = 62.0 - 7.0 = 55.0$ mm (from baseline)
Bottom $Y = 20.0 + 7.0 = 27.0$ mm

Note: Slots and keyways are often dimensioned to the centerlines of the end semicircles. If this is the case, it is only necessary to change these to baseline dimensions.

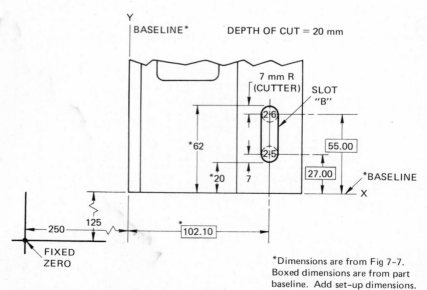

Fig. 7-20   Dimensioned sketch for milling keyway B.

The three depth cams are set so that they will stop their rapid downward motion at 5.0 mm above the top of the part, or 5.0 mm above the bottom of the previous cut. The first cam thus travels $5.0 + 8.0 = 13$ mm at slow vertical feed rates. The next two cams will each descend at rapid feed, 8.0 mm further, then 13 mm more at slow feed.

The program for this cut is given in Fig. 7-21.

Explanation of Fig. 7-21

n035
    g80   Rapid to programmed X and Y, etc., as in previous programs.
    m56   First cam cuts 8.0 mm deep

n036
    g79   Quill descends at rapid and feed rate, cutting to 8.0 mm depth at bottom of slot.

n037 Y 18000   Causes the table to move at feed rate, to the top end of the slot.

n038
    g81   Cutter stays where it is in X and Y and goes up at a rapid rate so that the next depth/cam can be used.

n039
    g79   Calls for cam m57 which cuts 8.0 mm deeper than m56.

n040–n043   Repeat the above cycles.

n044
    g81   Raises the cutter to home position at rapid traverse.

OPERATION	SEQ. NO.	PREP FUNCT.	X POSITION	Y POSITION	MISC. FUNCT.	POS. NO.	REMARKS
MILL SLOT "B"	n035	g80	x10210	y15200	m56	25	RAPID TO POS.
(SEE FIG. 7-20)	n036	g79				25	LOWER QUILL
	n037			y18000		26	
	n038	g81				26	
	n039	g79			m57	26	NEW DEPTH
	n040			y15200		25	
	n041	g81				25	
	n042	g79			m58	25	FINAL DEPTH CUT THRU
	n043			y18000		26	
	n044	g81				26	RAISE QUILL
	n045	g80	x52400	y10000	m02		OFF WORK END OF JOB

Fig. 7-21 WORD ADDRESS program for milling keyway *B*.

n045

g80 The table moves to the new (X, Y) location without any vertical movement of the spindle. The X52400 leaves the spindle about 50 mm to the right of the workpiece so that the part may easily be replaced. Y10000 leaves the cutter near the front of the table so that the operator can easily change the cutter.

m02 End of the program. Machine stops and the tape rewinds to beginning of first block, n001.

The entire program is shown in Fig. 7-22.

The rpm and feed rates are calculated as shown below, assuming that the cutting speed is 120 m/min, and feed rates are as shown.

Formulas used are:

$$\text{rpm} = \frac{300 \times \text{cutting speed}}{\text{cutter diameter}}$$

X and Y axis feed (mm/min) = (rpm)(feed per tooth)(number of teeth)

Z axis (vertical feed) = about $1/4$ the above.

N/C PROGRAMMING SHEET – WORD ADDRESS

PART NO. FIG.7-1,7-7 TO 7-20	PART NAME ADAPTER PLATE	MTL. ALUM. 6061-T6	PAGE 1	PAGES OF 3	DATE		BY			
OPERATION	SEQ. NO.	PREP FUNCT.	X POSITION	Y POSITION	MISC. FUNCT.	POS. NO.	SPEED RPM	FEED mm/min SPNDL	FEED mm/min TABLE	REMARKS
SET UP FRONT L.H. CORNER OF PART AT X250.00, Y125.00										
FACE MILL "F"	n001	g78	x36110	y10774	m51	1	1125	110	450	CLAMP QUILL
32 mm - 4 FLUTE END	n002	g79		y30226		2				
MILL. FIG. 7-8	n003		x34438			3				
	n004			y10774		4				
	n005	g78				4				UNCLAMP QUILL
	n006	g81				4				
FACE MILL "A"	n007	g79	x31838	y10774	m52	5				POSITION X,Y,Z
FIG. 7-10	n008			y30226		6				
	n009		x29495			7				
	n010			y10774		8				
	n011		x27152			9				
	n012			y30226		10				
	n013	g81				10				RAISE QUILL
CUT NOTCH "E"	n014	g78	x24252	y10774	m51	11				POS. X & Y & Z CLAMP QUILL
FIG. 7-12	n015	g79		y30226		12				
	n016	g78				12				UNCLAMP
	n017	g81			m06	12				TOOL CHANGE

Fig. 7-22 (This page and facing page): Complete WORD ADDRESS program for machining the part shown in Fig. 7-7. All dimensions and feed rates in metric.

N/C PROGRAMMING SHEET – WORD ADDRESS

PART NO. FIG. 7-1, 7-7 TO 7-20	PART NAME ADAPTER PLATE	MTL. ALUM. 6061-T6		PAGE 2	PAGES OF 3	DATE		BY		
OPERATION	SEQ. NO.	PREP FUNCT.	X POSITION	Y POSITION	MISC. FUNCT.	POS. NO.	SPEED RPM	FEED mm/min SPNDL	FEED mm/min TABLE	REMARKS
MILL EDGE "C"	n018	g80	x38660	y11124	m53	13	1440	80	230	POSITION X, Y
FIG. 7-14	n019	g79				13				POSITION Z
25 mm - 2 FLUTE	n020			y29876		14				
END MILL	n021	g81				14				RAISE QUILL
ROUGH MILL	n022	g80	x30190	y22090	m54	15				POSITION X, Y
POCKET "D"	n023	g79				15				POSITION Z
FIG. 7-16	n024			y25780		16				
	n025		x28400			17				
	n026			y22090		18				
	n027	g81			m06	18				RAISE QUILL TOOL CHANGE
FINISH MILL POCKET "D"	n028	g79	x27670	y21360	m55	19	3000	80	300	POSITION X, Y, Z
12 mm - 2 FLUTE END MILL	n029		x30920			20				
FIG. 7-18	n030			y26510		21				
	n031		x27670			22				
	n032			y21360		23				
	n033		x30000	y24000		24				TOOL INTO CLEAR
	n034	g81	m06			24				TOOL CHANGE
MILL SLOT "B"	n035	g80	x10210	y15200	m56	25	2570	40	154	POSITION X & Y
14 mm - 2 FLUTE END MILL	n036	g79				25				POSITION Z
	n037			y18000		26				

PART NO. FIG. 7-1, 7-7 TO 7-20	PART NAME ADAPTER PLATE	MTL. ALUM. 6061-T6		PAGE 3	PAGES OF 3	DATE		BY		
OPERATION	SEQ. NO.	PREP FUNCT.	X POSITION	Y POSITION	MISC. FUNCT.	POS. NO.	SPEED RPM	FEED mm/min SPNDL	FEED mm/min TABLE	REMARKS
MILL SLOT "B"	n038	g81				26				
CONTINUED	n039	g79			m57	26				NEW DEPTH
	n040			y15200		25				
	n041	g81				25				
	n042	g79			m58	25				CUT THRU
	n043			y18000		26				
	n044	g81				26				
	n045	g80	x52400	y10000	m02					CUTTER OFF THE WORK. STOP MACHINE. REWIND TAPE.
— RELEASE PART & PUT IN TOTE BOX										
— SECURE NEXT PIECE										
— CHANGE TOOL TO 32 mm MILL										
— PUSH "CYCLE START" AND MACHINE										

Calculations:

32-mm-dia. cutter (4-flute)

$$\text{rpm} = \frac{(300)(120)}{32} = 1125 \text{ rpm}$$

Feed (*X* and *Y* axes) = (1125)(0.1)(4) = 450 mm/min

25-mm-dia. 2-flute end mill

$$\text{rpm} = \frac{(300)(120)}{25} = 1440 \text{ rpm}$$

Use 0.08 mm/tooth feed.
Feed = (1440)(0.08)(2) = 230 mm/min

12-mm-dia. 2-flute end mill
Use 0.05 mm/tooth feed for finish cut.

$$\text{rpm} = \frac{(300)(120)}{12} = 3000 \text{ rpm}$$

Feed = (3000)(0.05) (2) = 300 mm/min

14-mm-dia. 2-flute end mill
Use 0.03 mm/tooth because of enclosed deep cut.

$$\text{rpm} = \frac{(300)(120)}{14} = 2570 \text{ rpm}$$

Feed = (2570)(0.03)(2) = 154 mm/min

This fairly simple program illustrates the most frequently used types of milling cuts. This same type of N/C machine is often equipped to cut slopes and arcs (linear and circular interpolation), which will be described in the chapter on lathes.

There are, of course, numerical control machines which will specify feeds, speeds, and *Z* axis distances by tape command, using additional letter addresses. These will be described in later chapters. However, the basic programming shown in this chapter is also used on many of the more complex N/C machines.

Chapter 8

N/C TURRET DRILLS

Many jobs done on vertical-spindle N/C machines require the use of several kinds and sizes of cutting tools. On the machines considered so far, this requires the operator to stop the machine and change the tools. A good operator makes the change very quickly. However, it does take time and does require that the machine operator be at the machine most of the time.

The N/C turret drill, with 6, 8, or occasionally 10 tools on the turret, will change tools according to a tape command, so that the operator is free to do other work. The tools used include drills, end mills, face mills, taps, counterbores, boring tools, and reamers.

Numerical control turret drills are made in models in which the table and carriage move in the X and Y directions, as in Fig. 8-1, or in larger sizes in which the table is fixed and the column and spindle move, as in Fig. 8-2. Both two- and three-axis control systems are available. Only the three-axis machines are discussed in this book.

These machines can be purchased with controls for use with either absolute or incremental programming systems. Some are equipped to accept both types of dimensioning. The EIA suggests a g90 code to specify absolute dimensioning and a g91 code for incremental dimensioning. In this chapter we use absolute dimensioning, as that is the only system our machine will accept.

Many N/C turret drills have standard or optional equipment which allows inch/metric switchable use with either a switch on the console or EIA code g70 for inch and g71 code for metric dimensions. Both linear in-

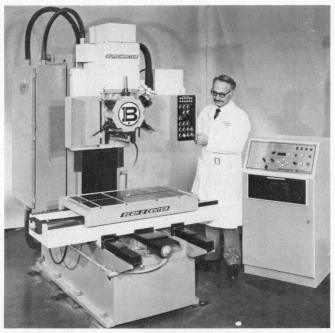

Fig. 8-1 An N/C turret drill, made with 6 or 8 tool turrets, and two or three axis controls. Available with inch/metric controls, linear and circular interpolation. Programming similar to that shown in this chapter. (*Burgmaster Division, Houdaille Industries, Inc.*)

terpolation and circular interpolation are also often available. These are discussed in Chap. 11.

The VARIABLE BLOCK WORD ADDRESS format is the most frequently used coding. Tab and code delete may be used for the convenience of the programmer and the tape-punch operator. The MCU disregards these two codes.

The new items introduced in this chapter on a three-axis N/C machine are the R and Z axes, some new g and m codes, and automatic tool-change coding.

Zero Location

Turret drills generally use a fixed-zero reference point. For our zero point we use the back left-hand corner of the machine table, a location used by some manufacturers, though other locations are also used. Even though this places all the point locations in the fourth quadrant, no minus signs

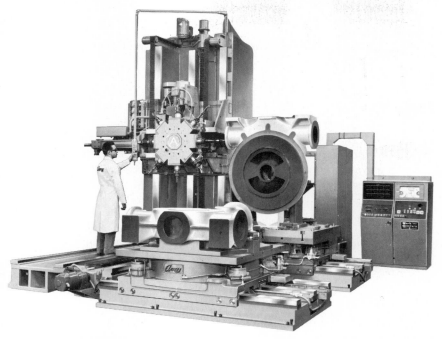

Fig. 8-2 A specially built eight-spindle N/C turret drill. WORD ADDRESS programming. This N/C machine has two "air lift" rotating tables, so that one can be loaded while machining is being done at the other. Travel is X, 4570 mm [180 in.]; Y, 1830 mm [72 in.] vertically; Z, 610 mm [24 in.] horizontally. (*Avey Machine Tool Co.*)

are used, since the MCU can measure the Y-axis distances in only one direction.

These machines are frequently supplied (as standard or optional equipment) with the FULL-RANGE ZERO SHIFT, which allows the operator to put the part anywhere on the machine table and bring the setup point X and Y settings up to the part. One method of accomplishing this is explained in Appendix C.

The Z Axis

An added advantage of the three-axis N/C turret drills is tape control of the vertical motion of the spindle—the Z-axis motion. Full control of this vertical motion requires control of four factors:

1. Distance from the tool point to the work at the start of each command
2. Distance through which the spindle moves at rapid advance

3. Distance through which the spindle moves at the tape-controlled feed rate
4. Dwell time at depth (if any used) and rate and direction of rotation of return stroke

There is no "standard" system of specifying control of these four items. However, most three-axis N/C turret drills use Z-axis control systems, which have the same basic elements as the method described in this chapter. We will consider these four factors one at a time.

1. *Tool-point-to-work distance* will vary considerably if the turret always starts from the same place. This is because different cutting tools will project different distances from the face of the turret, as shown in Figs. 8-1 and 8-2.

To correct this situation, a system called **tool-length compensation,** or **random-length tool setting,** may be supplied, or purchased as an added feature. This feature includes six or eight sets of dials or thumb wheels, one set for each turret station. Each set has five or six decade (meaning ten) switches which represent a depth dimension read to hundredths or thousandths of a millimetre.

Since there are several methods of using these tool-setting controls, the programmer must carefully read the manual for each N/C machine. However, there are certain similarities.

One way of using tool-length compensation is to set each cutting tool in turn against a shim of, say, 2.5-mm thickness. Each set of decade switches is adjusted so that the corresponding cutter will rapid-traverse

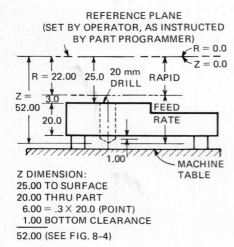

Z DIMENSION:
25.00 TO SURFACE
20.00 THRU PART
 6.00 = .3 × 20.0 (POINT)
 1.00 BOTTOM CLEARANCE
52.00 (SEE FIG. 8-4)

Fig. 8-3 Example of calculations of R and *Z* dimensions for one make of turret drill.

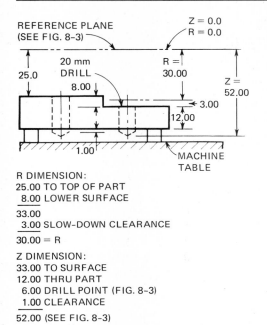

REFERENCE PLANE
(SEE FIG. 8-3)

Z = 0.0
R = 0.0

20 mm
DRILL
8.00

25.0

R = 30.00

Z = 52.00

3.00

12.00

1.00

MACHINE
TABLE

R DIMENSION:
25.00 TO TOP OF PART
 8.00 LOWER SURFACE

33.00
 3.00 SLOW–DOWN CLEARANCE

30.00 = R

Z DIMENSION:
33.00 TO SURFACE
12.00 THRU PART
 6.00 DRILL POINT (FIG. 8-3)
 1.00 CLEARANCE

52.00 (SEE FIG. 8-3)

Fig. 8-4   Calculations of R and Z for the lower level of the part shown in Fig. 8-3.

down to this preset point. The cutter will then continue downward at feed rate. In some control systems this becomes the Z zero point. In other systems, Z zero is at the actual part surface, usually the highest surface if there is more than one, and the 2.5 mm is automatically subtracted by the MCU.

In all cases, additional rapid vertical travel can be programmed by the use of an R function. The R represents *R*apid traverse, and in some cases it also establishes a *R*eference plane at a selected distance above the part surface.

A system used on many N/C turret drills, and the one we will use for our sample program, employs the random-length tool-setting feature to establish a Z zero reference plane above the workpiece. To do this, the part programmer decides that the minimum clearance between the points of the tools and the highest point on the workpiece should be 25 mm, or more if large clamps must be avoided. For our program we use a 25-mm clearance (Figs. 8-3 and 8-4).

When the operator is setting up the job, he or she places a 25-mm block on top of the highest point on the work (or at whatever location the N/C programmer has specified). The turret is then lowered, by jog control, so that the tool in turret position 1 just touches this block. Next the operator adjusts the set of dials numbered 1 so that a meter is nulled or made to read zero. Or sometimes an indicator light is used to indicate when the correct setting has been made.

This setting is left on the control console as long as this tool is used on this job. The rest of the tools are then **offset,** or adjusted to the same height, by the same method. Each turret station has its own set of dials; so each tool compensation setting may be different.

In some of the newer machine control units, up to 30 tool offsets can be preset at one time. With this MCU, an additional column on the program sheet is used to call out the offsets as d01 through d30. Under certain circumstances the operator can preset the offsets for several jobs, according to the programmer's list. Then, as each job is run, the operator merely loads the tools (preset to length), the tape, and the work, and starts the machine. The proper selection of tool offsets is specified on the tape, so the operator does not have to spend time on this part of the setup.

Notice that if tools become dull, and a new tool is placed in position, it can be set to the same nulled reading, or the dials may very quickly be readjusted for the new cutter length.

The foregoing procedure establishes a common Z and R zero plane 25 mm above the workpiece. Thus all Z-axis programming may be completed from a common plane, which greatly simplifies the programmer's work. The programmer must keep in mind when setting this plane that this is the highest the tool will go above the part, except when using g80 and TOOL CHANGE, as explained later.

2. *Additional rapid travel,* below the tool-setting distance just established, is specified as the R distance in this part program. We have used a 25-mm-setting distance; so a command of R022 (trailing zeros omitted) would specify rapid downward travel of 22.00 mm. This brings the tool point to 3.00 mm above the work. This 22.00 dimension is shown in Fig. 8-3, and a different R dimension for the lower surface is illustrated in Fig. 8-4. The clearance of 3.00 is necessary in order to let the vertical travel of the spindle slow down to the cutting feed rate before it touches the work. The minimum allowable clearance varies according to the N/C machine, and is specified by the manufacturer.

3. *Control of vertical feed rate distance* is accomplished by use of a specified Z distance. In the programming of the turret drill used in this chapter:

Z distance = distance from a preset point above workpiece
to the lowest point in the tool travel

Z = R + depth of cut + clearance + point allowance

(Clearance and point allowance are zero in most milling cuts.) The *difference* between the Z and the R dimensions is the distance during which cutting will be done at the programmed vertical feed rate. This difference is computed by the machine control unit and is illustrated in Figs. 8-3 and 8-4. Since all the tool points have been set to the same zero reference plane, the R and Z dimensions are easily used to specify rapid travel and feed rate travel for varying heights on the workpiece and various depths of cutting. A sample of the necessary computations is shown in Figs. 8-3 and 8-4.

4. *Control of dwell and rotation of spindle* is accomplished by the use of the proper g and m codes, as explained later in this chapter.

Coding Used (Word Address)

Many of the g and m codes used in turret drills are the same as those we have used in previous programs. However, in the constant effort to get faster, more flexible control systems, several additional g codes are used, especially if the N/C machine has all the available options. The EIA Standard RS-274-C has reserved g60 to g69 for these special codes, and g81 to g89 are for fixed cycles (or canned cycles). EIA formerly specified what these fixed cycles should be, and most manufacturers follow these earlier specifications fairly closely, though not completely. Codes used in this chapter which are not EIA-suggested standards are marked with an asterisk (*).

The coding to be used on our machine is as follows.

g (Preparatory) Function Codes

g80   CANCEL cycle. This cancels any previously programmed cycle. It is also used to prevent downward movement of the spindle and, when programmed with a tool change, to bring the turret back to home, or fully retracted position, instead of to the programmed R dimension.

g81   DRILLING cycle. This code will cause the table to move at rapid traverse in X and Y, and then the spindle will go down at rapid traverse to the programmed R dimension, then at the programmed feed rate to the Z dimension, and then retract at rapid traverse to the R dimension. Notice that this is the standard DRILL cycle, except that it saves time by retracting only to the R dimension, which is frequently only 2.5 mm above the surface of the part.

g82   DWELL cycle. The same as g81, except that it will cause the cutter to dwell at the bottom of the Z dimension. The length of the dwell is regulated by a manually set timer on the console. This is ordinarily used for spot facing and counterboring.

g84   TAPPING cycle. The same as g81, except that at the bottom of the Z dimension the spindle rotation is reversed, and it feeds up at the programmed feed rate to the programmed R dimension. Then it resumes clockwise rotation, and the next command is read from the tape.

g85   BORE cycle. The same as g81, except that at the bottom of the Z dimension the spindle is retracted **at feed rate** up to the R dimension. This cycle is also used when reversible tapping heads are used instead of the machine's g84 cycle.

*g86   MILLING cycle. Usually, when this code is used, the table has been moved to the X and Y of the first milling position by means of a g80 command. The g86 then causes the spindle to rapid-traverse down to the R dimension and feed to the Z depth. The spindle is then **automatically** clamped, and the next command is read from the tape.

The next command is usually a move in either the X or the Y direction, since this is a straight-cut system. This move will be at feed rate. This milling feed rate may be manually set, or (sometimes as an extra-cost option) it may be programmed on the tape.

The spindle is raised by using the g80 code.

Note: g86 as used here is not an EIA standard code. It corresponds to the g79 used in Chap. 7. The g86 is used in this program, since this coding is used on some N/C turret drills in industry. Neither EIA nor AIA suggests a specific coding for a milling operation. Thus several different code numbers are in use. The two milling codes mentioned above are reasonably widely used.

Other g codes are used for special purposes (often as extra-cost options), but those listed above are codes for the basic canned cycles (automatic complete cycles) used on turret drills.

m (Miscellaneous) Function Codes

m00   PROGRAM STOP. This code will cause the tape reader to stop, and thus allow time for the operator to perform some special job. This might be to gage a bored or tapped hole or to insert and use special tooling. The operator must press the *start* button to continue the machining.

*m04   INHIBIT CREEP, or CREEP OFF. This is not an EIA standard code, but it is a command which is used by several N/C machines, with various m code numbers. The m04 and m05 used here are not necessarily the coding used by any actual N/C machine.

Normally, when the machine is moving in the Z axis, it is automatically programmed (in the MCU), so that it will slow down to a creep for the last one or two millimetres of travel, and not overshoot the depth command.

The INHIBIT CREEP code cancels the slowdown and allows a few tenths of a millimetre overshoot. It is valuable when drilling "through" holes, in which a slight extra travel is not important. The time saved on each hole is not large, but in a day's run it can be worthwhile.

*m05 INHIBIT CREEP is turned off; so **creep to exact depth** is now in operation. This is needed only after using an m04 command.

m08 **Turns on the coolant.** In some N/C machines there is a choice of mist or flood coolant. In these machines, m07 is also a COOLANT ON code.

m09 **Turns off the coolant.** Will turn off either the m07 or the m08.

m30 END OF TAPE. Retract spindle, stop machine, rewind tape. Practically the same as the previously used m02. The EIA descriptions are almost identical, and they leave the choice to the manufacturer.

As in the case of the PREPARATORY functions, other m codes may be used, but those listed above constitute a basic list.

Feed and Speed

Some of the older N/C machines coded speeds and feeds with a three-digit coded number. This was called the "magic-three" code, and is explained in Appendix B. None of the newer machines use this coding.

Speed codes are preceded by the letter S. These may be coded directly in rpm, but few N/C machines have continuously variable speeds, so only certain speeds can be coded. Sometimes speeds are coded from a published table with a prearranged coding system, as shown in Table 8-1. This is the table used in this chapter. Two digits follow the S, and no decimal point is used.

Referring to Table 8-1, the first digit specifies the range, and the second digit is the speed column. Thus 465 rpm is coded s34.

Feed rates are specified by several different systems. Some older N/C machines multiplied the feed rate by two and rounded off the result. The EIA describes an "inverse time" coding which is used especially with linear and circular interpolation on a few machines.

Table 8-1 SPEED CODING FOR N/C TURRET DRILL USED IN CHAP. 8

Range	Speed, rpm					
	1	2	3	4	5	6
1	30	36	45	55	65	75
2	90	110	130	155	200	225
3	270	325	390	465	560	670
4	800	965	1160	1390	1670	2000

Today many numerical control machines have more sophisticated electronics, so they program the feed rate in mm/min (or ipm) directly, with no special coding. Some machines go one step further and permit feed rate coding in the basic quantity of mm/rev (or ipr). In this chapter we will code feed rates in mm/min, to the nearest 12 mm.

The range of feeds and the number of steps (or continuous variation) vary considerably in the N/C equipment available. Our machine has feeds available from 12 to 1200 mm/min in steps (increments) of 12 mm. Coding used in this chapter is three numbers with the decimal point to the right of the third number. Code letter is f. Thus a feed rate of 271 mm/min rounds off to $271 \div 12 = 22.58$. Use $22 \times 12 = $ f264.

Cutting Tools

An eight-station turret drill is being used, but we need only seven of these. The code letter is T, and two digits are used. The spindles holding the tools are numbered, and may be called for in any order. This is random tool selection. The cutting tools, tool numbers, speeds, and feeds are shown in Table 8-2.

Table 8-2   CUTTING-TOOL LIST FOR N/C TURRET DRILL

Tool no.	Description	Computed rpm	Actual rpm and code*	Computed feed, mm/min	Feed code
t01	Spot drill, 12-mm dia.	875	800 s41	64	f060
t02	6.9-mm drill	1522	1390 s44	111	f108
t03	10-mm drill	1050	965 s42	77	f072
t04	16-mm counterbore	469	465 s34	23	f024
t05	12-mm 82°	625	560 s35	28	f024
t06	M8 x 1.25 tap	300	225 s26	281	f336
t07	20-mm 4-flute end mill	375	325 s32	65	f252 f060
t08	Not used				

* rpm and code from Table 8-1.

Material is cold-rolled steel. Cutting tools are standard-grade high-speed steel. Cutting speeds used are drilling, 35 m/min; milling, counter-sink, and counterbore, 25 m/min; tapping, 8 m/min or less. Actual rpm used is the closest available from Table 8-1, which is less than that computed. Feed rates are rounded off to the nearest feed rate code number, as specified previously.

Omit Trailing Zeros

In Fig. 3-11, this method of shortening the coding of X, Y, and Z coordinates was described. This system is also used in this chapter for R and F quantities. To review briefly, quantities are written as follows:

$X = 12.000$ is coded as $X012$
$Y = 0.750$ is coded as $Y00075$
$Z = 1.300$ is coded as $Z0013$

Notice that **preceding** zeros must be used, and it is assumed that there are three digits to the left of the decimal point as we are using the metric system.

A PART PROGRAM

The Part Drawing

The part to be machined by our N/C turret drill is shown in Fig. 8-5. This drawing is dimensioned properly, from the viewpoint of the machine

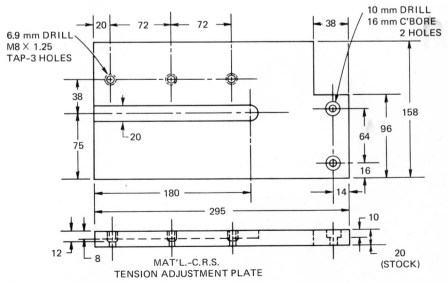

Fig. 8-5 Drawing of part to be programmed.

designer, but requires considerable arithmetic if it is to be used by an N/C part programmer. Therefore this has been redrawn in Fig. 8-6, showing all dimensions given from the setup point. This drawing also shows where the machine operator is to locate the part on the machine table. Of course, if the full-zero-adjust feature is available (Appendix C), the operator can locate the part wherever convenient. Figure 8-6 is the source for the dimensions on the manuscript.

Manuscript Form

Since the coding used is WORD ADDRESS VARIABLE BLOCK format, the exact arrangement of the column headings in Fig. 8-7 is not important. The arrangement shown is quite logical, and is the same as, or very similar to, that used on N/C turret-drill part-programming manuscripts in industry.

Discussion of Part Program (Fig. 8-7)

The values of the X and Y coordinates are easily computed by adding the setup dimensions (125, 250 mm) to the part dimensions shown in Fig. 8-6; so no discussion of most of these is necessary. Note that if a dimension re-

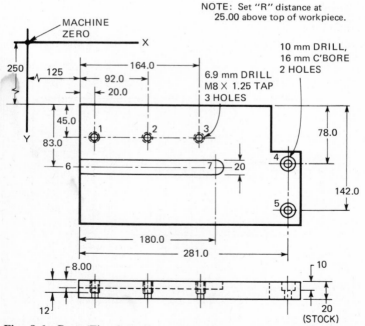

Fig. 8-6   Part (Fig. 8-5) shown located on machine table, and dimensions refigured for baseline dimensioning.

mains the same in the blocks that follow, it does not need to be written again. This is the same procedure used in previous programs.

n001

g80 — Used in this system wherever a tool change is being made. Some N/C turret drills use the m06 for tool change, in the same way it was used in Chaps. 6 and 7. In our machine, the X and Y positioning can be done during the tool changes (turret positioning) because the g80 brings the turret to "home" position and prevents any vertical movement. The desired tool is coded on the same line as the g80.

t01 — Cutting tool (spot drill) in the first turret station is brought into cutting position.

m08 — Coolant is turned on.

n002

g81 — Starts and completes the drilling cycle, which is also used for spot drilling.

r022 — The reference plane was set at 25 mm above the part, so this coding allows rapid downward travel for 22 mm (to within 3 mm of the part surface).

z0268 — The total depth to be drilled is:

25.00 tool tip to part surface

$\underline{1.80}$ = 0.3 × 6 = 1.8 mm depth to feed a 118° drill to

26.80   get a 6-mm-dia. spot drill

s41 — rpm for a 12-mm drill at 35 m/min cutting speed is (300)(35) ÷ 12 = 875 rpm

The speed table (Table 8-1) shows that s41 (800 rpm) is the closest speed which is less than computed.

f060 — Feed = (800 rpm)(0.80 mm/rev) = 64.0 mm/min. The 0.08 mm (0.003 in.) per revolution is arbitrarily decided by the programmer.

n003

Only the Y coordinate changes.

n004

Only the Y coordinate changes.

n005

Both X and Y change. All other columns remain the same.

n006

Only the Y coordinate changes.

n007

g80 — Required to prevent Z motion and bring the turret to home position. Return to hole 1 location, and change tool to a 6.9-mm tap drill (turret position 2). It would be just as good programming to drill the 10-mm holes next, or even to mill the slot first. Each programmer may decide on a different order of operations and be equally correct.

n008

g81 and r022   The same as n002.

z0481   Total depth of cut is

25.00 tool tip to part surface
20.00 part thickness
 2.10 = 0.3 × 6.9 drill point
 1.00 bottom clearance
48.10 total

s44   rpm = (300)(35) ÷ 6.9 = 1522 rpm
    In Table 8-1, use 1390 rpm, which is coded s44.

f108   Feed = (1390 rpm)(0.08 mm/rev) = 111.2 mm/mm. Closest multiple of 12 = 108

n009 and n010

Change in dimensions only.

n011

Move to hole 4 and change to 10-mm drill, turret position 3. See n007.

n012

z049   Total depth of cut is

25.00 tool tip to part surface
20.00 part thickness
 3.00 = 0.3 × 10 drill point length
 1.00 bottom breakthrough (clearance)
49.00 total depth

s42   rpm = (300)(35) ÷ 10 = 1050 rpm. From Table 8-1, use 965 rpm = s42.

f072   Feed code = (965)(0.08) = 77.2 mm/min. Closest multiple of 12 = 72

m04   Repeated after tool change. Creep off.

n103

Change in Y only.

n014   Tool change is the only command actually required since the counterbore will be done at the same position as in n013. However, the X and Y coordinates are usually repeated, so that a complete block of information is available in case start-up is needed at this block.

n015

g82   DWELL cycle for counterbore. Time is preset on the console.

PROGRAM SHEET – N/C TURRET DRILL

PART NO. *FIG. 8-5 & 8-6*			OPER. NO.		DATE			PAGE *1* OF *2*	
PART NAME *TENSION ADJ. PLATE*			TAPE NO.		MTL. *C.R.S.*			PROG. BY	

SEQ. n000	PREP g000	X x000.00	Y y000.00	R r000.0	Z z000.00	FEED f000	RPM s00	TOOL t00	MISC. m00	REMARKS
n001	g80	x145	y295					t01	m08	*POSITION @ HOLE #1, TOOL #1*
n002	g81			r022	z0268	f060	s41			*SPOT DRILL HOLE #1 TO 6mm DIA.*
n003		x217								*SPOT DRILL #2*
n004		x289								*" " #3*
n005		x406	y328							*" " #4*
n006			y392							*" " #5*
n007	g80	x145	y295					t02		*POSITION TO HOLE #1, CHANGE TOOL*
n008	g81			r022	z0481	f108	s44		m04	*DRILL #1 THRU CREEP OFF*
n009		x217								*TAP DRILL #2*
n010		x289								*TAP DRILL #3*
n011	g80	x406	y328					t03		*POSITION TO HOLE #4, TOOL #3*
n012	g81			r022	z049	f072	s42		m04	*DRILL #4 THRU CREEP STILL OFF*
n013			y392							*DRILL #5*
n014	g80	x406	y392					t04		*CHANGE TOOL SAME POSITION*
n015	g82			r022	z035	f024	s34		m05	*C'BORE #5 CREEP ON*
n016			y328							*C'BORE #4*
n017	g80	x289	y295					t05		*POSITION TO #3 CHANGE TOOL*

PROGRAM SHEET – N/C TURRET DRILL

PART NO. *FIG. 8-5 & 8-6*			OPER. NO.		DATE			PAGE *2* OF *2*	
PART NAME *TENSION ADJ. PLATE*			TAPE NO.		MTL. *C.R.S.*			PROG. BY	

SEQ. n000	PREP g00	X x000.00	Y y000.00	R r000.0	Z z000.00	FEED f000	RPM s00	TOOL t00	MISC. m00	REMARKS
n018	g81			r022	z0308	f024	s35			*C'SINK TO 10mm DIA., HOLE #3*
n019		x217								*C'SK #2*
n020		x145								*C'SK #1*
n021	g80	x145	y295					t06		*CHANGE TOOL SAME POSITION*
n022	g84			r022	z04325	f336	s31			*TAP HOLE #1*
n023		x217								*" " #2*
n024		x289								*" " #3*
n025	g80	x113	y333					t07		*POSITION AT #6 CHANGE TOOL*
n026	g86			r03	z033	f252	s32			*DOWN TO DEPTH FOR MILLING SLOT*
n027		x305				f060				*MILL TO POSITION #7*
n028	g80	x	y					t01	m09 m30	*RETRACT SPINDLE, GO TO HOLE #1, REWIND*

Fig. 8-7 Program manuscript for machining the tension-adjusting plate (Figs. 8-5 and 8-6) on an N/C turret drill. Dimensions in millimetres.

z035   Counterbore depth is

25.00 tool tip to part surface
10.00 depth of cut
35.00 total depth

s34   rpm = (300)(25) ÷ 16 = 469 rpm. In Table 8-1, use 465
            rpm = s34.

f024   Feed code is (465)(0.05) = 23.3 mm/min. Use 24 mm/min,
            f024.

m05   Creep is on to maintain close depth control.

n016
Change in *Y* only.

n017   Position to hole 3, and change to 82° countersink, used to chamfer
before tapping. Creep stays on.

n018

z0308   Total depth of cut to make a 10-mm-dia. 82° countersink.
            The multiplier for an 82° point angle is 0.58 × diameter (Ap-
            pendix A, Table A-3).

25.00 tool tip to part surface.
 5.80 = .58 × 10
30.80 total depth

s35   Speed code = (300)(25) ÷ 12 = 625 rpm. Use 560 rpm,
            s35.

f024   Feed code = (560)(0.05) = 28 mm/min. Use 24 mm/min.

n019 and **n020**
Changes in *X* only.

n021   Stay at hole 1, and change tool to M8 × 1.25 tap. Repeat *X* and *Y*
for convenience.

n022

g84   Tapping cycle code.
r022   Same as before.
z04325   Total depth is

25.00 tool to work
12.00 depth of full thread
 6.25 = 5 × 1.25 for five partial threads
43.25 total depth

s31   Speed code = (300)(8) ÷ 8 = 300 rpm. Use 270 rpm in
            table.

f336   Feed rate = (270)(1.25) = 337.5 mm/min

Note: Tap feed = pitch in mm × rpm. Use 336 mm/min.

n023

Change in X only.

n024

Change in X only.

n025

Change to 20-mm-dia. end mill, and position the center of the cutter at point 6 on Fig. 8-6, clear of the workpiece.

$$\text{Point 6: } X = 125 - 10 \text{ tool radius} - 2 \text{ clearance}$$
$$= 125 - 12 = 113 \text{ mm}$$
$$Y = 250 + 83 = 333 \text{ mm}$$

n026

g86	MILLING cycle, brings cutter down to r and z depth at same location. Locks spindle and reads next block.
r030	Rapid down to 30 mm, as cutter is not on the work, and a few seconds are saved.
z033	Final depth is

25.00 tool tip to work
 8.00 depth of cut, from the drawing
33.00 mm total depth

s32	Speed = $(300)(25) \div 20 = 375$ rpm In Table 8-1, 325 rpm = s32
f252	Vertical feed. Fairly fast, as cutter is ''cutting air.''

n027 Move in X only to centerline of the end of the slot, position 7. Cutter stays locked in down position under control of the previous g86.

x305	$= 125 + 180 = 305$ mm
f060	Feed code = $(325)(4 \text{ flutes})(0.05 \text{ mm/tooth})$ $= 65$ mm/min. Use f60.

n028

g80	Releases spindle and raises it to home position when programmed with a tool change.
X and Y	With no number following them, in this N/C machine specifies $X = 0.0$, $Y = 0.0$. The table could be positioned to some other convenient location for loading and unloading the work.
t01	Bring tool 1 into position.
m09	COOLANT OFF.
m30	END OF TAPE, rewind, etc. Notice that both m codes may be punched into the tape in the same block. As these m codes are often switching commands, several N/C units allow this type of programming.

This example of programming has illustrated
1. A commonly used method of specifying tool numbers
2. The use of "omit trailing zeros"
3. One way of using Z and R notations
4. A modified method of specifying feed rates
5. A timesaving CREEP OFF code used on some N/C machines, both turret drill and single spindle

Chapter 9

FIXED BLOCK
N/C PROGRAMMING

All the programs considered so far have been VARIABLE BLOCK format. "Format" means the "form" or "arrangement" of the coding. The blocks (between the EOB codes) which were previously used may have included only one or two code "words" or may have been made up of several words. Thus the blocks varied in length. However, some numerical control machines use a coding system which requires that every block must be exactly the same specified length. This system is not used as frequently as the VARIABLE BLOCK, TAB SEQUENTIAL, or WORD ADDRESS. However, FIXED BLOCK programming is defined by EIA and is used by some well-known N/C machines and controls.

Fixed Block Format

FIXED BLOCK coding does not use either tab or letter codes to tell the machine control unit what signal is being sent. In the FIXED BLOCK, the codes for each section of the MCU must always be located in the same position in every block. The numerical control machines shown in Figs. 9-1 and 9-9 use this format for programming.

The length of the block will vary according to the number of functions which are to be controlled by tape commands. One system which is being used is shown in Fig. 9-3. This system uses a 20-digit block, which means that every block must have 20 rows, and every row, as shown in Fig. 9-3, has a special meaning to the MCU, so it must always contain a digit. This is the system which is used in this chapter.

Thus, even though an X or Y location does not change, it must be repeated in the next block. Theoretically, letters as well as numbers may be

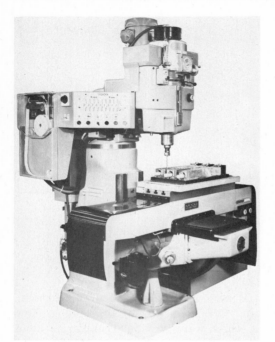

Fig. 9-1   A compact, 3-axis N/C drilling, boring, milling, and tapping machine. Uses FIXED BLOCK programming. Manually selected feeds and speeds, though can be tape-selected on other models. (*Moog Inc., Hydra-Point Division.*)

included in this format, though letters are not used in the example in this book.

Manuscript Form and Coding

One style of FIXED BLOCK N/C manuscript form is shown in Fig. 9-8. The headings are quite similar to those used in previous programs in this book. The form shown is ruled across so that the vertical space between lines is 12.7 mm [¹/₂ in.], which is correct for standard typewriter triple spacing. Thus the person punching the tape can conveniently type just below the handwritten codes, which simplifies checking the typeout. The explanation of the column headings is as follows:

Description column is used for remarks and hole location or cutter size and position-number notations. These the programmer has marked on the blueprint for his or her own convenience. The information in this column is not punched into the tape.

Sequence number is the same as in previous programs. It must have three digits. This sequence, or block number, is often shown on the console so that the machine operator can more easily follow the program.

Preparatory function is the same as the g coding shown in Chaps. 6 and 7. In our FIXED BLOCK program this column will not contain the letter g, and instead of the two-number g81, etc., only one row is allowed, so a special set of code numbers is used.

X and Y are both written with five digits, and in the metric mode the decimal point is understood to be after the third digit from the left. The decimal point is not punched into the tape.

Z feed point is coded in the X column and is called for by a 9 code in the PREP. FUNC. This is explained later.

Z final depth is coded in the Y column, after a 9 code, and is explained later.

Reserved function column is not used, so it is always coded with a zero.

Tool function may, as in the machine shown in Fig. 9-9, actually cause an automatic tool changer to select the next cutter. On other models it may turn on a light on a table with number tool storage positions, or it may merely start a buzzer which alerts the operator that a new tool is needed. This requires two digits.

In more advanced N/C machines, this column, working with the MISC column, specifies speeds and feeds from preestablished tables.

Miscellaneous function, as in previous programs, specifies stops and tool changes. This function is allowed two rows, so EIA coding is often used, but without the letter m.

In the more sophisticated machines this column is also used to specify speed and feed rate change axes and other special functions, and works with the special TOOL FUNC. CODING.

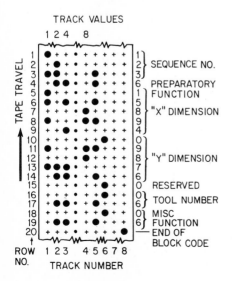

Fig. 9-2 A standard block of tape for one FIXED BLOCK format. (*Moog Inc., Hydra-Point Division.*)

The PREPARATORY and MISCELLANEOUS codes which we are most interested in are as follows. This is not a complete list, but it is all we need for our problem.

Preparatory Functions

0 code   POSITION ONLY. Table moves to programmed X and Y, but with no Z motion, the same as g80 previously used.

1 code   TAPPING. Table moves rapidly to X and Y. Feed and speed must be coordinated to get proper pitch, and an axial floating tap holder must be used. Spindle rapids and feeds down, reverses rotation and feeds and rapids up to home position.

5 code   DRILLING. Table moves in rapid traverse to programmed X and Y. Rapids down to feed point (see Fig. 9-5), proceeds down at the preset feed rate to final depth, and retracts to home position. If an 81 MISC command is used, it retracts only to the feed point.

6 code   MILLING. Spindle moves to final depth and then, at set milling feed rate, to the programmed X and Y in that block of tape.

7 code   BORING. Same as 5 code, but spindle retracts at feed rate to the feed point and then retracts rapidly.

8 code   TAPPING—PITCH CONTROL. On some models, a special lead screw can be put in the machine. This accurately controls the pitch of the thread. Otherwise, it is the same as code 1.

9 code   READ Z BLOCK. The dimensions in the X and Y columns, after this 9 PREP. FUNC., become the depth dimensions for the preceding block. This depth information is then used on all future operations until a new 9 code command is given. See Fig. 9-5 for computing these dimensions.

Miscellaneous Functions

00   PROGRAM STOP. After completing all commands in the tape block, the machine stops, coolant flow stops. Push cycle *start* to continue.

02   END OF PROGRAM. Machine and coolant flow are stopped. Tape rewinds if rewind option is used.

06   TOOL CHANGE (manual). Machine and coolant stop after completing commands in the block. A buzzer signals the operator, and in some models a light is lit at the proper station on the tool storage table.

07   FLOOD COOLANT ON.

08   MIST COOLANT ON.

09   COOLANT OFF. Not needed after an 00, 02, or 06 command.

80   NO CHANGE IN MISCELLANEOUS FUNCTIONS. This maintains the MISC circuit previously specified.

81   PARTIAL SPINDLE RETRACT ON. This limits spindle retraction to the feed point. Cannot be used with PREPARATORY FUNCTIONS 1 and 8

but saves time when drilling. Watch out for interference of high points on the part or the clamping.

82 PARTIAL SPINDLE RETRACT OFF. Resume standard spindle retraction to home spindle zero position.

FEED POINT and FINAL DEPTH calculations are made according to Fig. 9-5. The knee of the machine is raised, usually by hand, to allow an inch or so clearance when the longest tool to be used is in the fully retracted (home) position.

The FEED POINT is often 0.50 to 2.00 mm. above the work surface. The spindle comes down at rapid feed to this point.

The FINAL DEPTH is the full depth of stroke, to the point of the tool. The spindle travels at the programmed feed rate the distance between these two points. This is, on Fig. 9-5, the FEATURE DEPTH plus the CLEARANCE.

A Fixed Block Program

For ease of programming, a combined part and location drawing is shown on Fig. 9-7. The dimensions were taken from a part drawing and converted to coordinate dimensions as shown.

Notice that the electronic fixed zero is at the back left-hand corner of the machine table. Because the Y distances are all in one direction from zero, they are programmed as positive numbers, even though they are in the third quadrant.

Machine Description

The N/C machine used in this program has the following specifications:
1. Single vertical spindle with quick-change toolholders.
2. Table moves in X and Y by tape control.
3. Spindle moves in Z axis to programmed depths, using preset tool lengths (Fig. 9-6).
4. Speeds and feeds are changed manually by the operator.
5. Tool number codes turn on buzzer and turn on lights on a tool storage table.
6. Buffer storage, which means that while performing the instructions in one block of tape, the next block is being read into the MCU. This makes much faster response to the coding.
7. Zero adjust of 1.02 mm [0.040 in.] is available in both X and Y axes.

Tooling

Since this is a very simple shape, it is decided to use a milling vise with a special corner jaw, as shown in Fig. 9-3. This illustration also shows one

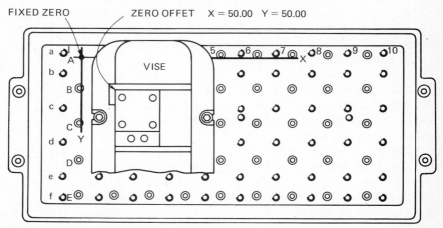

FIXED ZERO   ZERO OFFET   X = 50.00   Y = 50.00

Fig. 9-3 Predrilled subplate on machine table, and vise with angle jaw holding part at correct setup location. (*Moog Inc., Hydra-Point Division.*)

type of "subplate" which may be attached to the machine table. This one has jig-bored holes with bushings for dowels, and tapped holes for hold-down bolts for use with a vise (as shown) or with fixtures or with clamps. Subplates are used with many N/C machine setups. Often the user makes up his or her own plate, sometimes drilling, reaming, and tapping it on the N/C machine itself.

Fig. 9-4 An assortment of tools and toolholders used on N/C machines, and a simple, accurate tool-setting device. (*Moog Inc., Hydra-Point Division.*)

Cutting tools must be set to a specified length if the Z axis is tape-controlled. Figure 9-6 shows a simple method of accurately setting the tool length. The drawing of Fig. 9-5 shows the information needed in order to calculate the feed-point and final-depth Z dimensions. The work height is usually the highest point on the workpiece, as shown on the blueprint. Feature depth is also taken from the print.

It is advisable to tabulate the tool numbers, lengths, etc., in some standard form. One simple method is shown in Fig. 9-6. This form, or another similar to it, is of considerable help in computing the Z dimensions, and it also provides a record so that the job can easily be set up the next time it is run.

The **actual program** for milling the 25-mm-wide by 12.5-mm-deep flat surface and drilling six 10-mm holes on Fig. 9-7 is shown in Fig. 9-8. We

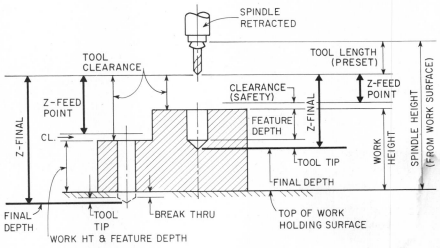

Tool clearance is set so that the longest tool being used will clear the vise (or the clamps) and the workpiece by 1 in. or as specified by the N/C programmer.

Feed-point dimension = FP

FP = spindle height − work height − tool length − clearance

or

FP = tool clearance − safety clearance

final depth dimension Z = ZF

ZF = FP + clearance + feature depth + tool tip + break through

(Use tool tip if needed. Tool tip is zero for most milling cutters. Break through is zero for blind-drilled holes and some milling cuts.)

Fig. 9-5   Method of calculating the Z dimensions for the N/C machine used in Chap. 9.

TOOL NO.	DESCRIPTION	DIAMETER	TOOL LENGTH	TOOL CLEARANCE	TOOL TIP	SPEED RPM (m/min)	FEED mm/min (mm/rev.)
01	END MILL	32.00	125.0	37.50	0	234(25)	X,Y=60.0(0.26)
02	DRILL	10.00	137.50	25.00*	3.0	1050(35)	Z=106 (0.1)
03	TAP DRILL	5.10	75.00	87.50	1.54	2050(35)	Z=102(0.05)
04	TAP-M6-1	6.00	100.00	62.50	5.00	200(4)	Z=200(1.0)
	ALL DIMENSIONS IN mm						

*Minimum clearance for the longest tool. Clearance of shorter tools is correspondingly greater. Figured from "Home" (fully retracted) position.

Fig. 9-6 Tool list for part shown in Fig. 9-7.

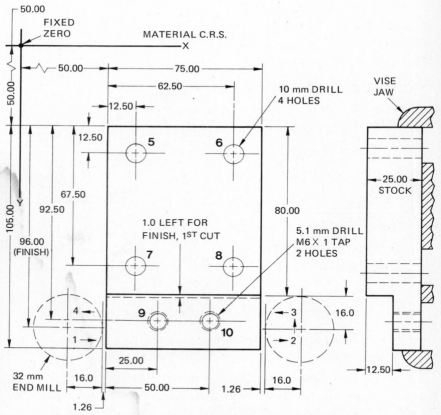

Fig. 9-7 Combined part location, hole numbering, and milling-cutter location drawing.

are assuming that the four edges have previously been machined. As the material is 25 × 75 mm [1 × 3 in.] cold rolled steel (CRS), the bottom and top surfaces will be smooth as received. A brief explanation of the principal features follows:

Seq. no.

001 Standard LOAD, UNLOAD, TOOL CHANGE block. *X* and *Y* are specified to locate the table at a point which is convenient for the operator for changing the workpiece and the tool.

   PREP. FUNC. Zero prevents any vertical motion.

002 ROUGH MILL. Locates the cutter just off the part at position 1, with 1.00 mm left for a finish cut to the 80.00 dimension.

$$X = 50.00 - 16.0 \text{ cutter radius} - 1.26 \text{ clearance}$$
$$= 50.00 - 17.26 = 32.74 \text{ mm}$$
$$Y = 50.00 + 80.00 \text{ to edge} + 16.00 \text{ cutter radius} + 1.00$$
$$\text{left for finish cut.}$$
$$= 130.00 + 17.00 = 147.00 \text{ mm}$$

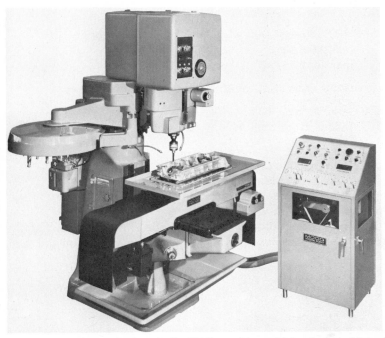

Fig. 9-9 A 3-axis single-spindle N/C machine which can use 24 tools with its random access tool changer. Manual Data Input (MDI), and tapping with or without lead screw. Capacity 22 mm [⁷/₈ in.] mild steel. (*Moog Inc., Hydra-Point Division.*)

PREP. FUNC. 0 for positioning only
003   Lower cutter and move to position 2

X = 50.00 + 75.00 part width + 16.00 cutter radius
    + 1.26 clearance
    = 125 + 17.26 = 142.26 mm
Y = same as previous.

PREP. FUNC. 6 is the milling command. It first lowers the spindle to depth and then moves to the programmed position. Notice that this is opposite to the usual action of first locating and then going to depth. 80 in MISC column continues the 07 command.

004   DEPTH COMMAND. Preparatory function 9. This 9 function sends the following 10 digits to the Z axis control section of the machine control unit. Some other N/C machines use this method of utilizing the same columns for two different actions.

This depth command continues to control all depths until it is changed by a new function 9 command.

FEED POINT (Figs. 9-5 and 9-6) must be at least 1.0 mm less than final depth.

FP = 37.50 tool clearance + 12.50 feature depth − 1.0
    = 49.00 mm
ZF (final depth) = 49.00 + 1.00 = 50.00 mm

005   Moves cutter to start the finish cut at position 3.

X = same as Seq. 003.
Y = 147.00 (from Seq. 003) − 1.00 (left for finish)
    = 146.00 mm

All other commands remain the same.
006   Moves cutter to position 4, completing the finish cut.
007   TOOL CHANGE, at the same position as block 001. The 06 MISC FUNC. turns off the coolant and the spindle.
008   DRILL HOLE 8; this could have been any one of the six drilled holes.
PREP. FUNC. 5 first positions at the programmed location and then drills the hole.

X = 50.00 + 62.50 = 112.50 mm
Y = 50.00 + 67.50 = 117.50 mm

07 COOLANT ON again.
009   DEPTH OF DRILL, PREP. FUNC. 9

FP = 25.00 tool clearance − 1.00 safety clearance = 24.00 mm
ZF = 25.00 tool clearance + 25.00 feature depth
    + (0.3)(10.00) drill point + 1.00 breakthrough
    = 50.00 + 3.0 + 1.0 = 54.00 mm

MATERIAL _____ C.R.S. _____

PART NO. _____ FIG. 9-7 _____ REV. 2 _____

PART NAME _____

ZERO OFFSET X _____ 50.00 mm _____ Y _____ 50.00 mm

PREPARED BY _____ DATE _____

TYPED BY _____ DATE _____

DESCRIP TION	SEQ. NO.	PREP. FCT.	FEED POINT X-AXIS 000.00 OR	FINAL DEP. Y-AXIS 000.00	RES. FCT.	TOOL FCT. CODE	MISC. FCT.
LOAD CHG.TL. POSIT.	001	0	17500	00000	0	00	02
MILL TO 1	002	0	03274	14700	0	01	07
MILL TO 2	003	6	14226	14700	0	01	80
DEPTH	004	9	04900	05000	0	01	80
MOVE TO 3	005	6	14226	14600	0	01	80
MILL TO 4	006	6	04900	14600	0	01	80
TOOL CHG.	007	0	17500	20000	0	02	06
DRILL 8	008	5	11250	11750	0	02	07
DEPTH 5 TO 8	009	9	02400	05400	0	02	81
DRILL 7	010	5	06250	11750	0	02	80
DRILL 6	011	5	11250	06250	0	02	80
DRILL 5	012	5	06250	06250	0	02	80
TOOL CHG.	013	0	17500	20000	0	03	06
DRILL 9	014	5	07500	14250	0	03	07

DESCRIP TION	SEQ. NO.	PREP. FCT.	FEED POINT X-AXIS 000.00 OR	FINAL DEP. Y-AXIS 000.00	RES. FCT.	TOOL FCT. CODE	MISC. FCT.
DEPTH 9 & 10	015	9	09900	11504	0	03	81
DRILL 10	016	5	10000	14250	0	03	80
TOOL CHG.	017	0	17500	20000	0	04	06
TAP 10	018	1	10000	14250	0	04	07
DEPTH TO TAP	019	9	07400	09350	0	04	80
TAP 9	020	1	07500	14250	0	04	80
UNLOAD POSITION	021	0	17500	00000	0	00	02
- UNLOAD FINISHED PART							
- LOAD NEXT PIECE							
- CHANGE TO 10mm DRILL							
- PUSH CYCLE START TO CONTINUE							

Fig. 9-8 Program manuscript for machining Fig. 9-7 on a machine with FIXED BLOCK programming.

143

81   MISC function stops the drill retraction at the FP instead of it going all the way up to home position.

010–012   DRILL holes 7, 6, 5. The X and Y coordinates are easily figured from Fig. 9-4.

013   TOOL CHANGE. See Seq. 007.

014   TAP DRILL hole 9, at different depths from previous hole.

X and Y are easily figured from Fig. 9-7. The rest of the coding is the same as in Seq. 008.

015   DEPTH commands (Figs. 9-5 and 9-6)

FP = 87.50 tool clearance + 12.50 feature depth
    − 1.00 clearance = 99.00 mm
ZF = 87.50 + 25.00 feature depth + 1.54 drill point
    + 1.00 breakthrough = 115.04 mm

016   TAP DRILL hole 10.

X and Y from Fig. 9-7.

017   TOOL CHANGE, as before.

018   TAP hole 10. Same X and Y as Seq. 016.

019   DEPTH commands

FP = 62.50 + 12.50 to surface − 1.00 clearance = 74.00 mm
ZF = 62.50 + 25.00 feature depth
    + 5.00 (for 5 incomplete threads) + 1.00 breakthrough
    = 93.50 mm

020   TAP hole 9. X and Y as before.

021   UNLOAD, same as Seq. 001.

02   MISC FUNC. Stops the tape, the spindle, and the coolant.

Note: The 81 (partial retract) function must be used with care, so that the spindle does not bump into clamps or higher parts of the workpiece.

Notice that the actual programming of this type of machine is very similar to that of the WORD ADDRESS programs of Chaps. 6 and 7. The printout of the tape for Seq. 002 would be:

0020032741470000107(E0B)

This looks difficult to read, but a little experience makes it easy to separate the group of numbers. This same machine can also ream, bore, countersink, and counterbore.

Chapter 10

AUTOMATIC TOOL CHANGERS AND A HORIZONTAL-SPINDLE THREE-AXIS N/C MACHINE

The N/C turret drills discussed in Chap. 8 illustrate a simple form of automatic tool changing which handles 6 to 10 cutting tools, each on its own spindle. There is also a group of **single-spindle** numerical control machines which provide automatic tool changing. These machines have a tool magazine which holds a number of cutting tools. The number of tools varies: 15-, 30-, 39-, and 60-tool-capacity magazines are in common use.

Types of Tool Changers

There are five principal variations in the design of these automatic tool changers:

1. The tool magazine may have a vertical or horizontal axis. Usually, the axis of the tool magazine is in the same direction as the spindle axis. Thus, in Fig. 10-1, both axes are vertical, and in Figs. 10-2 and 2-10, both are horizontal. However, a few machines, such as those shown in Figs. 10-3 and 1-3, have the tool-magazine axis at an angle to the spindle axis.

2. The tool magazine may rotate so that the center of the tool is automatically aligned with the spindle, as in Fig. 10-1. More frequently, the tool magazine is at one side, or above the spindle, as shown in Figs. 2-10 and 10-2. As can be seen in the illustrations, the mechanism for each tool changer is quite different. In fact, in Fig. 2-10 this is quite a large piece of equipment.

3. The tools may be selected sequentially or at random. Some tool changers require that the tools be used in order—numbers 1, 2, 3, . . . , in succession. Random tool selection is more frequently

Fig. 10-1 (*a*) Close view of a vertical-axis rotary-tool magazine. Notice that each pocket is permanently numbered. Notice also the wide variety of tools that can be used. (*Cleereman Machine Tool Corp.*)

used, which means that tools may be used in any order—numbers 1, 5, 12, 3, . . . , as desired. The numbers refer to the number of the position, or pocket, on the tool magazine. These, as shown in Fig. 10-1, are clearly marked, and the tape is coded to call for a certain pocket.

This means that the part programmer must decide which pockets to use and supply the setup person with a list showing which cutting tool is to be stored in each position. Of course, if certain sizes of drills, taps, or reamers are frequently used, they can be assigned permanent positions in the magazine, leaving the rest of the pockets for special tools for each job.

4. For some N/C machines, such as those shown in Figs. 10-2 and 2-10, the tools themselves are given numbers, so that they may be placed at random in the tool magazine. The tool numbers, in two popular systems, are in the same BCD code as the holes in punched tape.

Fig. 10-1 (*b*) A horizontal-axis tool changer for a vertical-axis N/C machining center. Uses a single gripper, but still exchanges tools in less than ten seconds. (*Burgmaster Division, Houdaille Industries, Inc.*)

One method, shown in Fig. 10-4, uses a "key" in a special holder located in the tool magazine at the same station as the required toolholder. By cutting off, in a simple hand machine, all except the needed lugs, the tool number is represented in BCD.

Another method, shown in Fig. 10-5, uses rings and spacers along the toolholder shank for coding BCD tool numbers.

The coding system shown in Fig. 10-5*a* was the first one used, and it was very successful and sufficient for the tooling in use for many years. Its advantage is that, with a simple plastic "reader" like the one shown under the tool, an operator can, by checking the coded tool against the drawing of the same number, be certain

Fig. 10-2 One of the first machining centers made, and many are still working well today. Three-axis TAB SEQUENTIAL PROGRAMMING from a fixed zero. Uses a 7.5-kW [10-hp] motor, 457-mm [18-in.] square table. Full tape control of all functions, including random selections of up to 30 tools. (*Kearney & Trecker Corp.*)

Fig. 10-3 A horizontal-spindle three-axis machining center. Full floating zero, TAB SEQUENTIAL or WORD ADDRESS programming. Random tool selection, and contouring in *X-Y* axis. (*Colt Industries, Pratt & Whitney Machine Tool Division.*)

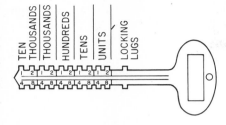

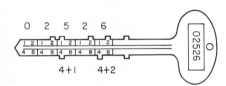

Fig. 10-4 A key used for coding tool numbers on some N/C machines. Top illustration shows the key as received. Bottom illustration shows how the key would look when cut to code tool 02526. (*Scully International, Inc., Bendix Corp.*)

that the tool is the correct type. He or she can also check all the dimensions before putting the cutter into the machine's magazine.

The "zero" is part of the tool holder, so it can be placed in a setup block of known height, on a surface plate, so the tool length can easily be checked. One simple system for checking tool lengths is shown in Fig. 9-4.

The newer set of rings shown in Fig. 10-5*b* allows a larger assortment of tools to be individually specified. It is "read" and set to length in the same way as just described.

In both types, a reading head close to the revolving tool magazine "feels" for these coded numbers until it finds the tool number which matches with the tool number specified on the tape. When the two numbers match, the proper cutter is positioned so that it can be picked up by the tool-changing mechanism.

5. Since about 1972 a "chain" type toolholder has become popular in both American and European N/C machining centers. Figure 10-6 shows one of these machines. This configuration enables the designer to get many tools in a relatively small space. The "chain" may be at either side or on the top of the N/C machine. Usually random tool selection is used, with the next tool selected and positioned while the last tool is cutting.

Tool-Size Limits

The magazines for most types of automatic tool changers have holes around the perimeter of a circle or on a chain, spaced at a certain distance. Of course, the diameter of any cutter cannot be greater than the hole spacing if all pockets are to be filled. Thus the maximum diameter of a cutter held in the magazine of one N/C machine is 70 mm [2³/₄ in.], another is 67 mm [2⁵/₈ in.], and another 101 mm [4 in.]. Cutters over these limiting sizes must be manually loaded into the machine spindle. This is one reason for the m00 PROGRAMMED STOP command. Some tool magazines will allow larger diameters if three pocket spaces are used for one tool. One machine specifies a 279 mm [11 in.] maximum diameter for a cutter such as a face mill.

Most tool-changing equipment will handle only a limited length of cutter. This varies considerably. Three N/C machines specify 190, 457, and 610 mm [7¹/₂, 18, and 24 in.] maximum lengths. Beyond these limits manual loading is required.

When specifying the use of the longer tools, the programmer must be especially careful that the workpiece is moved far enough away from the spindle, so that no collision will take place when the long tool is being moved into the spindle. Often the "home" position of the spindle is planned to avoid any possible collision. This can be done by placing the work or the fixture on the worktable so that the Z distance is as large as possible.

Preset Tool Lengths

The length just mentioned is the distance which the toolholder plus the cutting tool projects beyond some fixed point. This fixed point is usually

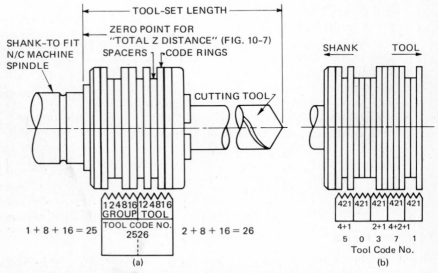

Fig. 10-5 Two methods of coding tool numbers by the use of rings and spacers positioned on the toolholder. Notice that Z zero is also on the toolholders. (*Kearney & Trecker Corp.*)

Fig. 10-6   A modern three-axis N/C machining center. Will contour in any two axes at the same time. Made with tables 1220 × 915 mm [48 × 36 in.] and up to double this size. WORD ADDRESS programming, random tool selection, 18.6-kW [25-hp] motor. Uses "chain type" tool holder. (*Kearney & Trecker Corp.*)

the face of the machine spindle, which is often (in this type of N/C machine) also the zero point for the Z axis.

As shown in Fig. 10-7, this tool length, or setting length, subtracted from the total Z distance (less any clearance), is the programmed Z dimension. Thus, if a dull drill, or end mill, etc., is replaced, the setting distance, or tool length, of the new tool must be the same as the original, or the actual depth of cut will be changed.

This is not too difficult, since there are several simple mechanical and optical tool-setting devices. In fact, some shops make their own. Sometimes the tools and toolholders are set by the toolroom, and sometimes the machine operator, having the time and the ability, makes up preset tools so that they will be ready when needed.

Many shops maintain a set of tool-setting drawings. These drawings describe the cutter, give it a tool number, and show the setting distance. Then, every time a 13.5 standard drill is needed, the part programmer can check the drawing and find, for example, that a 13.5-mm drill is tool

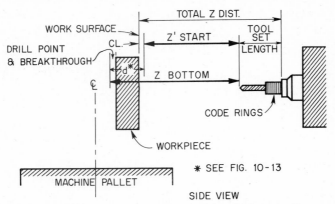

Fig. 10-7 Diagram for computing programmed *Z* distances as shown in Fig. 10-13.

number 1375 and has a setting distance of 140 mm. With this information the part programmer can write a program and feel confident that everyone concerned is talking about the same dimensions (see Fig. 6-5).

Other Machine Features

The tool-changing mechanisms, as can be seen in the illustrations mentioned previously, vary greatly. There are, however, two basic systems. If a cutter is selected by the pocket number in the tool magazine, as in Fig. 10-1, it must be returned to this same pocket. The tool changer must then get out of the way while the next tool selected is brought into position. The changer then grips this tool and places it in the spindle. This is **sequential** tool selection.

If the desired cutter is selected by its own assigned number, as in Figs. 10-2, 10-3, and 10-6, it can be placed anywhere in the tool magazine. Then, during a machining cycle, the magazine is rotated until the desired tool is in position. When the actual change is called for, the tool changer removes the previous tool, puts the next tool in the spindle, and puts the previous tool back in the magazine in the position just made empty. In both systems, the total time is usually from six to fifteen seconds. This is **random** tool selection.

The worktables also are available in a wide variety of sizes and shapes. The standard worktable for two of these larger N/C machines is a 457-mm [18-in.] square pallet. This pallet can be equipped so that it will rotate to any one of eight positions (every 45°) by tape command, or to every 0.001°, which allows 360 000 positions; and one machine even has a 1 000 000-position table available.

The number of axes under full tape control varies from two to five. The fourth axis is sometimes the ability to rotate a square or circular table at a controlled feed rate while cutting a specified path. The fifth axis may be to tilt either the worktable or the spindle assembly at a controlled feed rate while cutting. Any two, and sometimes three, of these axes may be moving at controlled feed rate at the same time.

The rated horsepower varies from 3.7 to 30 kW [5 to 40 hp] or over. The speeds available range from as low as 10 rpm to as high as 4000 rpm, though not on any one machine. The feed rates on the various axes may be varied from about 13 to 1270 mm/min [0.5 to 50 ipm] with rapid traverse up to 5000 mm/min [200 ipm].

The programming system used depends on the make of control system specified by the buyer. Many of these machines can be furnished with any one of several makes of machine control units, and each, of course, is programmed somewhat differently. Both WORD ADDRESS and TAB SEQUENTIAL systems are used.

Feeds and speeds formerly were coded by the magic-three method (Appendix B). Speeds also may be specified by referring to a coded table of available speeds or by programming the actual speed desired. We here program the actual speed, to the nearest 10 rpm. Feeds may be programmed from a coded table, or the code may be computed by several different systems. We program the actual feed rate, to the nearest 2.5 mm/min.

PREPARATORY and MISCELLANEOUS **functions** and the words in WORD ADDRESS programs are, in most of these N/C machines, quite close to EIA and AIA standards. Some machines use different coding for special functions. Our machine uses the EIA standard numbers **without the letter** before the number.

AN N/C PROGRAM WITH TOOL CHANGER

As an example of an actual program, we select from all the variables just discussed one set of specifications for our N/C machine. The machine specified here is quite similar to the one shown in Fig. 10-2.

Specifications—Machine and Programming

Three axes NPC fully tape-controlled. The X, Y, and Z dimensions are given as six digits; the decimal point is understood after the third digit from the left when using millimetres for dimensions.

Absolute dimensioning is used.

Single horizontal spindle.

Machine table 457 × 457 mm, with eight positions, specified by tape, using one number. Table positions are shown in Fig. 10-8.

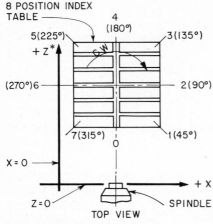

8 POSITION INDEX
TABLE

TOP VIEW

*OPPOSITE TO EIA STANDARD Z
NOTATION

Fig. 10-8 Diagram of the location of position numbers on the eight-position indexing worktable used in this chapter.

Fixed zero, as shown in Fig. 10-9.

Spindle speed 100 to 4000 rpm, specified as a four-digit number, to the nearest 10 rpm; i.e., 1457 rpm is coded as S1460 or S146 (drop trailing zeros).

Feed rates 12 to 750 mm/min, with rapid traverse at 3800 mm/min. Specified as a three-digit number with decimal point understood after the third number from the left, except rapid traverse, which is coded f8.

7.5-kW [10-hp] motor on spindle.

Tool coding system according to Fig. 10-5a.

Programming is TAB SEQUENTIAL; omit trailing zeros.

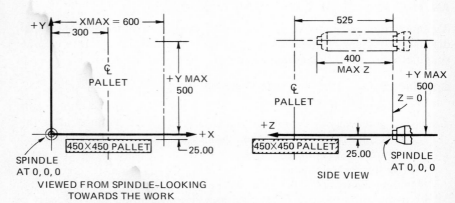

Fig. 10-9 Dimensioning system, and maximum dimensions of X, Y, and Z, for the N/C machine used in this chapter.

Miscellaneous Functions

These are the EIA standard m functions, but without the letter code, as follows:

00 PROGRAMMED STOP, for manual tool change or inspection
01 OPTIONAL STOP
02 END OF PROGRAM
03 CW SPINDLE ROTATION (as seen looking from the spindle toward the work)
04 CCW SPINDLE ROTATION
05 SPINDLE STOP
06 TOOL CHANGE (tool selection must be done in a previous block)
07 MIST COOLANT ON
08 FLOOD COOLANT ON
09 COOLANT OFF

Preparatory Functions (g codes)

These are not used. There are no canned cycles on this NPC machine, so every movement of the machine must be specified on the tape. This is the case with some other tape-controlled machines.

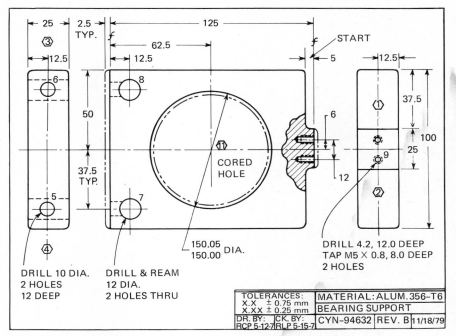

Fig. 10-10 Drawing of the part to be programmed in this chapter.

The Workpiece and Its Location

The part to be machined is shown in Fig. 10-10. The workpiece is made of cast aluminum, and two surfaces are to be milled, plus two drilled holes, two reamed holes, two tapped holes, and a bored hole. It would be wise to spot-drill all six holes. However, the tolerances are liberal: notice the tolerances in the title block. So, to shorten our sample problem, we omit the spot drilling.

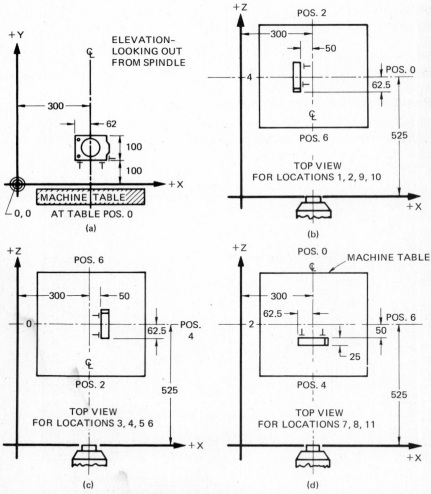

Fig. 10-11 Location drawings for computing X, Y, and Z dimensions for machining the part shown in Fig. 10-10.

One advantage of a horizontal-spindle N/C machine is that, with our rotary table, four to eight sides of a workpiece can easily be presented at right angles to the spindle, for easy machining access. Of course, the holding fixture often gets in the way, but usually at least three surfaces can be machined with one setup. The top of a workpiece (ninth side) can also be machined, using the periphery of a milling cutter.

This part is mounted as shown in Fig. 10-11, which allows access to the three surfaces we need to reach. Clamping is on the top and bottom. The fixture is secured to an angle plate on the machine table so that the part is located as shown in Fig. 10-11 in relation to the machine's fixed zero point.

Programming Forms

It often will save time if, before the actual programming is started, all the information on speeds, feeds, hole locations, etc., is assembled in an orderly arrangement, using forms similar to those of Figs. 10-12 and 10-13.

Sequence of operations and tool description sheet (Fig. 10-12) is a form which organizes all the necessary information on machining sequence, plus tool numbers, speeds, etc. Having this information tabulated will greatly simplify either manual or computer programming of the part.

SEQUENCE OF OPERATIONS AND TOOL DESCRIPTION SHEET								
PART NAME *BEARING SUPPORT*		DATE	MATERIAL *356-T6 Al.*	PAGE OF *1 1*	OPER. NO. *10*	PART NO. *CYN-94632*		
NO.	OPERATION DESCRIPTION	TOOL DESCRIPTION	TOOL DIAMETER	SETTING DISTANCE	TOOL CODE	SPEED RPM	FEED mm/min.	SPEED m/min.
1	*MILL TO 5 mm START DIM.*	*3 FL. END MILL*	*38 mm*	*150 mm*	*2101*	*1060*	*106*	*120*
2	*MILL 125 O.A.L.*	*SAME*						
*3	*SPOT DRILL (6) HOLES*	*118° SPOT DRILL*	*16*	*125*	*0102*	*1920*	*96*	*90*
4	*DRILL 4.2 (2) HOLES*	*DRILL-HSS*	*4.2*	*125*	*0103*	*3000*	*150*	*(36)*
5	*TAP M5 (2) HOLES*	*TAP-HSS*	*5×0.8*	*165*	*3104*	*240*	*192*	*(3.5)*
6	*DRILL 10 (2) HOLES*	*DRILL-HSS*	*10*	*175*	*0109*	*3000*	*240*	*90*
7	*DRILL 11.5 (2) HOLES*	*DRILL-HSS*	*11.5*	*175*	*0107*	*2480*	*158*	*90*
8	*REAM 12 (2) HOLES*	*REAM-STRT.FL.*	*12*	*190*	*0108*	*1200*	*150*	*45*
9	*RGH. BORE 149 mm DIA.*	*BORING BAR*	*149*	*250*	*0110*	*400*	*200*	*90*
10	*FIN. BORE 150.02 DIA.*	*BORING BAR*	*150.02*	*250*	*0111*	*400*	*50*	*90*
* SPOT DRILLING OMITTED FROM SAMPLE PROGRAM								

Fig. 10-12 One style of tabulation to expedite the part programmer's work involving tool information.

POSITION NO.	TABLE POSITION	X COORDINATE	Y COORDINATE	Z COORDINATE (TO PART SURFACE)	TOOL SETTING DISTANCE*	Z' START**	Z BOTTOM Z' + dt
1	6	237.5	187	462.5	150	312.5	———
2	6	237.5	113	462.5	150	312.5	———
3	2	362.5	222	462.5	150	312.5	———
4	2	362.5	78	462.5	150	312.5	———
5	2	362.5	112.5	462.5	175	285	305
6	2	362.5	187.5	462.5	175	285	305
7 Dr.	4	250	112.5	450	175	272.5	306
7 Rm.	4	250	112.5	450	190	257.5	290.5
8 Dr.	4	250	187.5	450	175	272.5	306
8 Rm.	4	250	187.5	450	190	257.5	290.5
9 Dr.	6	237.5	144	462.5	125	335	353.3
9 Tap	6	237.5	144	462.5	165	295	410
10 Dr.	6	237.5	156	462.5	125	335	353.3
10 Tap	6	237.5	156	462.5	165	295	410
11	4	300	150	450	250	197.5	227.5

* USING 2.5 mm CLEARANCE ALLOWANCE.
** Z' START = Z − TOOL-SET DISTANCE − CLEARANCE (IF ANY). SEE FIG. 10-7
† d = CLEARANCE USED IN Z' + DEPTH OF CUT + POINT LENGTH + BREAKTHROUGH.

Fig. 10-13 Coordinate location sheet. One style of tabulation which fixes all X, Y, and Z locations before the actual programming is started.

Coordinate location sheet (Fig. 10-13) shows the X, Y, and Z locations at each point in the program. It is usually easier and more accurate if the programmer concentrates on these locations as a separate job. Then, when he or she is writing the manuscript it will be possible to concentrate on the proper coding arrangement without stopping to compute values at each step.

The first column of the coordinate location sheet refers to numbers which the part programmer has assigned to each hole location and milling-cutter start-and-finish location. These numbers should be marked on the blueprint in colored ink or colored pencil.

The X, Y, and Z coordinate columns for our job are the absolute distances figured from the zero location—a fixed zero, in our example. Location sketches such as shown in Fig. 10-11 are very helpful, especially for the beginner.

Computing the X, Y, and Z coordinates for the drilled, tapped, and bored holes can be quite simply done by using the part drawing Fig. 10-10 and the location drawing Fig. 10-11. For example, the tapped holes, locations 9 and 10, are computed using Fig. 10-11a and b and the part drawing Fig. 10-10.

Holes 9 and 10:

$$
\begin{array}{rll}
X = & 300 \text{ mm} & \text{to centerline (Fig. 10-11}b) \\
& -50 & \text{to back of part} \\
& \underline{-12.5} & \text{to centerline of part} \\
& 237.5 & = X \text{ (also for hole 10)} \\
Y = & 100 \text{ mm} & \text{to bottom of part (Fig. 10-11}a) \\
& +50 & \text{to centerline of part (Fig. 10-10)} \\
& \underline{-6} & \text{to hole 9} \\
& 144 & = Y \text{ for hole 9} \\
& \underline{+12} & \text{to hole 10} \\
& 156 & = Y \text{ for hole 10} \\
Z = & 525 \text{ mm} & \text{to centerline of part (Fig. 10-11}b) \\
& \underline{-62.5} & \text{centerline to face of part} \\
& 462.5 & = Z \text{ coordinate (to part surface)} \\
& & \text{for both holes 9 and 10}
\end{array}
$$

The student can check the other hole locations in a similar manner.

Milling locations 1 to 4 (Fig. 10-13) are most easily located by using sketches such as those shown in Fig. 10-14. This is the same method used in Chap. 7. Often it is simpler to work from the centerlines, as shown in the following computations.

Locations 1 and 2 (Milling) (Fig. 10-14a)

Y coordinate (vertical):

$$
\begin{array}{rl}
100 \text{ mm} & \text{baseline to bottom of the part} \\
\underline{+50} & \text{bottom to centerline of part} \\
150 \text{ mm} & \text{baseline to centerline of part}
\end{array}
$$

The 38-mm-dia. cutter starts and ends at the same distance from the centerline. This distance is:

$$
\begin{array}{rl}
12.5 \text{ mm} & \text{centerline to edge of boss} \\
2.5 & \text{clearance} \\
\underline{19.0} & \text{milling cutter radius} \\
34.0 \text{ mm} & \text{center of part to center of cutter}
\end{array}
$$

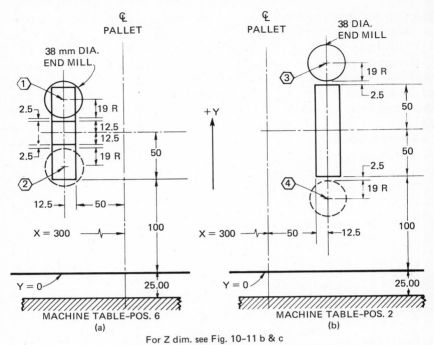

For Z dim. see Fig. 10-11 b & c

Fig. 10-14 Elevation of part (looking outward, from the spindle toward the work) showing X and Y locations for milling cuts. (a) Machine table—position 6, for Z dimensions (see also Fig. 10-11); (b) machine table—position 2, for Z dimension (see also Fig. 10-11c).

Using these two total figures:

Point 1: $Y = 150 + 34 = 184$ mm
Point 2: $Y = 150 - 34 = 116$ mm

X coordinate (horizontal) for locations 1 and 2:

300 mm	from X zero to center of pallet (Fig. 10-14a)
-50	center of pallet to face of part
-12.5	face to center of part
237.5 mm	$= X$ coordinate for locations 1 and 2

Locations 3 and 4 (Milling)

The student can easily verify the X and Y dimensions of these two locations, as shown in Fig. 10-13, by following the method shown above and using the drawing of Fig. 10-14b. Actually, an experienced part programmer can often compute these simple dimensions without a detailed sketch. It is still advisable to record all figures on a work sheet for future reference or debugging.

Tool-set distance (Fig. 10-13) is copied from the setting distance on the tool-description sheet (Fig. 10-12). The "Z start" column is computed by subtracting the "tool-set distance" and clearance from the "Z coordinate." Except with a facing cut, when the end mill may position to depth while it is off the work (e.g., points 1 to 4), a clearance must be subtracted to be certain that the tool point does not collide with the work. Our part is a small aluminum casting and usually has quite a good surface. Thus we allow only 2.5 mm clearance. However, when machining sand castings or forgings, it is not at all unusual to allow 6 mm or more, since a collision while the tool is moving at rapid traverse can cause very expensive damage to both the work and N/C machines. The formula for the "Z' start" is

$$Z' = Z - \text{set distance} - \text{clearance (if needed)} \qquad \text{Fig. 10-7}$$

The "Z bottom" column in our tabulation represents the final depth to which the tool point must move. This is figured by adding the total depth of cut d to the Z' figure. The total depth d must include, when applicable, the drill-point length, tapered threads on a tap, the clearance allowed in figure Z', plus breakthrough clearance. Thus

$$d = \text{clearance used in } Z' + \text{depth of cut} + \text{point allowance}$$
$$+ \text{breakthrough allowance}$$

The d figures used to compute the "Z bottom" dimensions in Fig. 10-13 are as follows:

Locations 1–4

Facing cuts start off the work, so go to full depth. No change from Z.

Locations 5 and 6

10-mm drill point = $10 \times 0.3 = 3$ mm
$d = 2.5$ clearance + 3.0 point + 12.5 depth = 18 mm = d

Locations 7 and 8

11.5-mm drill point = $11.5 \times 0.3 = 3.45$ mm
$d = 2.5$ clearance + 3.45 point + 25.0 depth
  + 1.5 bottom clearance = 32.45 mm = d (use 32.5)

Locations 7 and 8

12.0 ream (0.8 mm is standard 45° end chamfer).
$d = 2.5$ clearance + 0.8 chamfer + 25.0 depth
  + 1.5 bottom clearance = 29.8 mm = d

Locations 9 and 10

4.2-mm drill × 0.3 = 1.3-mm drill point
d = 2.5 clearance + 1.3 point + 12.0 depth
　= 15.8 mm = d

Locations 9 and 10

M5 × 0.8 tap
(Assume four-thread taper = 0.8 × 4 = 3.2 mm)
d = 2.5 clearance + 3.2 taper + 8.0 depth = 13.7 mm = d

Location 11

Rough and finish bore—usually no point allowance on boring bars.

d = 2.5 clearance + 25.00 depth + 2.5 bottom clearance
　= 30.0 mm = d

The program manuscript (Fig. 10-15) is for a TAB SEQUENTIAL program. For this reason, the order in which the columns are arranged is **fixed** by the requirements of the MCU. Remember, if a number in the MISC function column is the only item in a block, it must be preceded by **eight** tab codes. The same manuscript form, but with added columns for PREPARATORY commands, feed rate numbers, and arc commands, may be used for NCC and NPC-NCC machines.

THE PART PROGRAM

The actual writing of the program can now be done quite easily. Keep in mind that the programmer must, for this type of machine, program every single move that the machine makes—even, as you will see, to calling for a CCW spindle motion to retract the tap from the hole.

This means that the manuscript, or process planning sheet, will have many blocks of information, but not many characters per block. The part program is shown on Fig. 10-15.

Explanation of Part Program (Fig. 10-15)

General Information

1. There must be a tab character before the X coordinate and a tab for each succeeding function, whether or not it is used.
2. EOB may be punched after the last numerical information in the block. For example, if only a Z command were programmed, an EOB should be punched after the last Z-dimension digit.
3. A tool number should be called for one or more blocks before the tool change is to be made, but not the first line after an 06 tool

PROCESS PLANNING SHEET – HORIZONTAL SPINDLE N/C MACHINE									
PART NAME *BEARING SUPPORT*		MATERIAL *356-T6*		PAGE *1*	OF *3*	OPER. NO. *10*		PART NO. *CYN-94632B*	
SEQ. NO.	OPERATION	LONG X	VERT. Y	DEPTH Z	FEED RATE	TABLE INDEX	SPINDLE SPEED	TOOL CODE NO.	MISC. FUNCT.
1	START, POS. X & Y AT 1	2375	187	0	8	6	0	2101	
2	CHANGE TOOL								06
3	START SPINDLE - CW	2375	187	0	8		106		03
4	POS. Z COOLANT ON			3125					08
5	FACE TO 2		113		096				
6	RAISE Z			0	8				
7	INDEX TABLE					2			
8	POS. X & Y AT 3	3625	222						
9	POS. Z SELECT TOOL			3125				0103	
10	FACE TO 4		078		096				
11	RETRACT Z STOP SPINDLE			0	8		0		05
12	THE NEXT STEPS WOULD BE TO SPOT DRILL 6 HOLES, WHICH WILL BE OMITTED IN THIS PROGRAM. LAST OPERATION WOULD BE AT TABLE POSITION NO. 6								
13	POS. X & Y - #9 INDEX, TOOL	2375	144	0	8	6			06
14	START, POS. Z			335			3		03
15	DRILL #9 SELECT TOOL			3533	15			3104	
16	RET. Z TO CLEAR			335	8				
17	POS X & Y TO #10	2375	156						
18	DRILL #10			3533	15				
19	RETRACT Z STOP SPINDLE			0	8		0		05
20	TOOL CHG.								06
21	POS. Z START SPINDLE	2375	156	295	8	6	024		03
22	TAP #10			410	192				
23	REVERSE & TAP OUT TO CLEAR			295					04
24	POS. X & Y TO #9		144		8				03

Fig. 10-15 (*Continued on next two pages.*)

PROCESS PLANNING SHEET – HORIZONTAL SPINDLE N/C MACHINE									
PART NAME *BEARING SUPPORT*		MATERIAL *356 - T6*		PAGE *2*	OF *3*	OPER. NO. *10*		PART NO. *CYN-94632B*	
SEQ. NO.	OPERATION	LONG X	VERT. Y	DEPTH Z	FEED RATE	TABLE INDEX	SPINDLE SPEED	TOOL CODE NO.	MISC. FUNCT.
25	TAP #9			410	192				
26	REVERSE & TAP OUT			295					04
27	RETURN Z TO HOME			0	8		0		05
28	X & Y TO #8 INDEX TABLE	250	1875			4			
29	CHANGE TOOL								06
30	POS. Z START SPINDLE			2725	8		248		03
31	DRILL #8			306	158				
32	RETRACT Z SELECT TOOL			2725	8			0108	
33	POS. X & Y - #7		1125						
34	DRILL #7			306					
35	RETRACT Z STOP SPINDLE			0	8		0		05
36	CHANGE TOOL								06
37	ADV. REAM START SPINDLE	250	1125	2575	8	4	12		03
38	REAM #7			2905	15				
39	RETRACT Z SELECT TOOL			2575	8			0109	
40	POS. TO #8		1875						
41	REAM #8			2905	15				
42	RET. Z HOME STOP SPINDLE			0	8		0		05
43	CHG. TOOL INDEX TABLE					2			06
44	POS. X & Y - #6 START SPINDLE	3625	1875	0	8		3		03
45	ADVANCE Z			285					
46	DRILL #6			305	24				
47	RETRACT Z			285					
48	POS. TO #5		1125						

Fig. 10-15   *(Cont.)*

PROCESS PLANNING SHEET – HORIZONTAL SPINDLE N/C MACHINE									
PART NAME BEARING SUPPORT		MATERIAL 356-T6		PAGE 3	OF 3	OPER. NO. 10	PART NO. CYN-94632B		
SEQ. NO.	OPERATION	LONG X	VERT. Y	DEPTH Z	FEED RATE	TABLE INDEX	SPINDLE SPEED	TOOL CODE NO.	MISC. FUNCT.
49	DRILL #5			305	24				
50	RET. Z, STOP			0	8		0		05
51	STOP - MANUAL TOOL #0110								00
52	POS. TO #11 INDEX TABLE	3	15	0	8	4			
53	ADV. Z, START			1975			04		03
54	RGH. BORE #11			2275	05				
55	RET. Z, STOP			0	8		0		05
56	MANUAL TOOL #0111								00
57	ADVANCE Z START			1975	8		04		03
58	FIN. BORE #11			2275	05				
59	TOOL OUT @ FEEDRATE			1975					
60	RET. Z, STOP			0	8		0		05
61	REPLACE TOOL #0109								00
62	INDEX COOLANT OFF					0			09
63	END OF JOB								02

Fig. 10-15   (*Cont.*)

change. This allows the tool to be selected and positioned at the ready station while other work is being done.

4. The spindle must always be at its home position ($Z = 0$) before the table is indexed and before a tool change is made.

5. All dimensions are assumed to be positive (from the fixed zero); so no plus or minus signs are used. The $+X$ and $+Y$ dimensions shown are in the correct direction, according to EIA, as noted in Chaps. 2 and 4. The $+Z$ direction on this machine is **toward** the work, which is the opposite of the EIA standard. Figures 10-9 and 10-11 show this clearly.

Note: This difference in naming the $+Z$ direction becomes very important when using the incremental dimensions, since the plus and minus signs would specify the opposite directions.

6. The following program is, of course, only one way of machining this workpiece. Other sequences could be used which would be equally correct.

Comments on Program (Fig. 10-15)

All dimensions, speeds, feeds, tool numbers, and table positions are copied from the previously prepared sequence of operations and tool-description sheet and the coordinate location sheet. Thus comments will be limited to other elements of the programming.

Seq. 1 Locates spindle in X and Y for first operation at rapid traverse, places first tool in ready position, and positions Z at home.

Seq. 2 Places correct cutter in spindle.

Seq. 3 Repeats coordinates for convenience; starts spindle CW (03) at 1060 rpm.

Seq. 4 Moves spindle (Z) into position at rapid traverse (8) and turns on flood coolant (08).

Seq. 5 Moves down in Y at feed rate (106) to face off small boss.

Seq. 6 Returns spindle to $Z = 0$, at rapid traverse, so table may be indexed.

Seqs. 7–10 The same, except that the new tool 0103 (4.2-mm drill) must be brought to ready position. We chose to do this at sequence 9, though sequences 5 to 8 would be as satisfactory.

Seq. 11 Retracts Z to home (zero) position, ready for a tool change, and stops spindle rotation.

Seqs. 13–16 Typical of a drilling sequence with this kind of N/C machine. Remember to start the spindle (03) in sequence 14.

The **drilling sequence** could be stated as follows:

1. Position X and Y at rapid traverse; start spindle CW.
2. Rapid in Z close to part surface.
3. Drill to Z depth at feed rate.
4. Rapid out of hole, Z dimension, to clear work.
5. Spindle home and stop rotation (05), last hole only.

If a series of holes are to be drilled with the same drill, step 2 is eliminated, because step 4 brings the drill out only far enough to clear the work.

Seqs. 17–19 Illustrate the elimination of step 2, when the second hole is drilled. At sequence 19 spindle rotation is stopped, and spindle is brought home for a tool change.

Seqs. 21–24 Illustrate a tapping operation. At sequence 21 the table stays at position 10, so that no X or Y movements are needed. The full tapping sequences could be stated as follows:

1. Position X and Y at rapid traverse; start spindle in CW rotation (03), unless left-hand thread is wanted.
2. Rapid in Z to clearance dimension (first hole only).
3. Tap at feed rate.
4. Reverse spindle rotation (04 for right-hand thread) and retract tap from hole to Z clearance dimension.
5. Spindle home, and stop rotation (05); last hole only.

At sequence 21 a complete line of information is programmed, so that, if trouble develops, a start can be made, with all information on the tape. This is usually done every few blocks, especially after a tool change, not only in this type of program but in most N/C programming.

Seqs. 25–27   Repeats the tapping sequence, with no Z rapid.

Seq. 28   Indexes table to position 4 (180°) and positions spindle at hole 8.

Seqs. 29–35   Basically the same as the previous drilling at sequences 13 to 19.

Seqs. 37–50   Repetitions of the same type of cycle, used for reaming and drilling.

Seq. 51   Specifies PROGRAMMED STOP (00), sometimes called BLOCK STOP. This stops all further machine or tape motion until the operator presses his *cycle start* button.

At this time the operator will manually remove tool number 0109 and insert the 250-mm-long boring bar, set to 149 mm diameter.

Seqs. 52–55   Rough-boring cycle, which is the same as the drilling cycle. In sequence 55 the cutter is backed out at rapid traverse, with the spindle stopped.

Seq. 56   Another (00) block stop so that the finish boring bar, set to 150.02 mm diameter, may be manually put into the spindle.

Seqs. 57–58   Boring cycle.

Seq. 59   Brings boring bar back through the work **at feed rate** because tolerance is very close, and to avoid scratches or grooves.

Seq. 60   Returns spindle home and stops rotation.

Seq. 61   Block stop so tool number 0109 can be returned to the spindle. It will be put back in the tool magazine at the beginning of the next part, at sequence 2.

Seq. 62   Brings the table to position 0, so that the workpiece can be changed. On some machines the pallet is moved over, and another pallet, with a workpiece already secured, is moved into place. The second workpiece may be different from the first, and a second tape is used.

Seq. 63   END OF TAPE. Some machines will rewind the tape automatically, and some machines use a tumble box, which lets the tape hang down in folds. With the tumble box, the operator either uses a loop tape or rethreads the tape each time.

Conclusion

This method of programming is typical of several N/C machines. Some are TAB SEQUENTIAL, and some are WORD ADDRESS. The actual appearance of the tape for sequences 3 to 5 in Fig. 10-15 for TAB SEQUENTIAL would be as follows (* means tab symbol):

```
3*2375*187*0*8**106**06EOB
4***3125*****08EOB
5**113**096EOB
```

If the same program were written in WORD ADDRESS, it would be as follows:

```
n003x2375y187z0f8s106m03EB
n004z3125m08EB
n005y113f096EB
```

Tab symbols may be used, for convenient typeout of the WORD ADDRESS program, if desired. The letters, or words, might be typed as either capital or small (lowercase) letters, according to the equipment used for typeout. The computer uses only capital letters, while the tape typewriters are equipped either way, depending on the make and model.

Programs of this type take a large number of steps, or blocks of information. However, these steps follow simple patterns, so that they are not difficult to program or to follow. If the tool description and coordinate location sheets are carefully filled in using sketches such as Fig. 10-11 and numbered locations, the actual manuscript can be written quite quickly and accurately.

Chapter 11

ENGINE-LATHE
N/C PROGRAMMING

The study of the numerical control engine lathe (NCL) and numerical control turret lathe (NCTL) is especially valuable since these are simple contouring machines: ''simple'' because they use only two axes for most work, and in many machines the physical appearance and basic motions are very similar to those of a conventional lathe.

In this chapter we consider only the engine lathe, since the NCTL uses most of the same commands as the engine lathes, plus the planning and programming of the turret tooling. The NCTL can be learned on the job once the basic ideas of lathe programming are understood.

Figure 11-1 shows a view of the tooling on a numerical control turret lathe, and Fig. 11-2 illustrates a typical numerical control engine lathe. In Figs. 1-5 and 1-6 we saw other examples. Notice that these lathes are used for either turning between centers or for face-plate and chuck work. Table 11-1 shows the specifications of the engine lathe for this chapter and gives some idea of other specifications. The terms used are explained later.

Recent Advances

In the last few years many refinements have been made in the construction of and controls for N/C lathes. Some of these are:

1. An inch/metric switch is available on almost all models.
2. Either absolute or incremental programming is available, usually by using the proper tape code.
3. Feed rate is now specified on the tape in inches (or mm) per minute, and sometimes even in inches (or mm) per revolution. The complicated ''feed rate number'' has been eliminated.

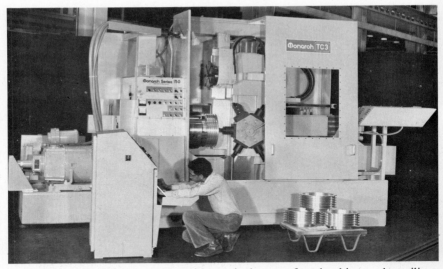

Fig. 11-1 An N/C turret lathe with vertical turret for chucking and auxiliary turret for turning. Available inch/metric and absolute/incremental WORD ADDRESS programming. Size shown has 450-mm [18-in.] chuck and up to 37-kW [50-hp] motor. (*Monarch Machine Tool Company.*)

Table 11-1 SPECIFICATIONS OF N/C LATHES

Item	English	Metric
Horsepower	20 to 75 hp	15 to 56 kW
Spindle speed	15 to 2000 rpm	same
Number of speeds	36 to 60	same
Feed rate	0.01 to 200 ipm	0.1 to 5000 mm/min
Rapid traverse	200 ipm	5000 mm/min
Maximum stock length	144 in.	3650 mm
Swing over bed	20 to 32 in.	500 to 800 mm
Maximum movement along X or Z, I and K, in one command	99.9999 in.	999.999 mm
I and K, thread leads	9.999 99 in.	99.9999 mm
Tool offsets	32 at 0.9999 in.	32 at 9.999 mm
Turret-cross slide	Hex, round, or square, 6 or 8 positions each.	

SOURCE: Adapted from the Monarch Machine Tool Company.

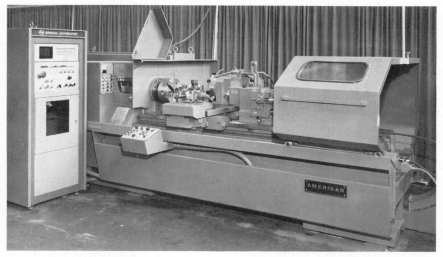

(a)

(b)

Fig. 11-2 (*a*) An N/C engine lathe shown with auxiliary rear turret. Front square or hexagonal turret can be used for chucking or between-centers turning. Inch/metric movements and feeds. Has 15-kW [20-hp] motor. Contouring controls standard, CNC available. (*b*) Close-up view of the eight-station optional rear turret. Used especially for turning and cutoff. (*American Tool Incorporated.*)

4. Slanted beds, chip conveyors, and chip removal equipment are available on many models.
5. Two turrets, for easier turning and face-plate work, are available on many models.
6. Tool offsets and maximum programmed distance in one block have been increased, making programming simpler.
7. "Check tapes" and built-in test panels make troubleshooting much easier.
8. CNC (see Chap. 13) is available as an extra on most lathes and is standard on some.
9. The use of "qualified" tooling saves setup time. These tools have a guaranteed ±.001 [0.02 mm] dimension from one side of the toolholder to the cutter location.

The Lathe Axes

As shown in Fig. 11-3, the N/C engine lathe uses only two axes, X and Z. This follows the basic rule that the Z axis is always the axis parallel to the centerline of the spindle. This means that the XY plane is in the same position as on any horizontal-spindle machine. Look back to Fig. 4-5, which illustrates this notation. However, the engine lathe usually has no way of moving the cutting tool up and down by tape control; so the Y axis is seldom used.

The numerical control turret lathe (NCTL) has a six-sided turret on a saddle, in addition to the one or two square turrets on the cross slide. This means that there are **two** Z-axis movements, which would be difficult to program. However, the EIA and AIA Standards provide for this. Their standard coding arrangement is shown in Table 11-2.

Thus the movement of the saddle turret is along the W axis, as shown in Fig. 11-4. Some turret lathes have a cross-sliding hex turret (parallel to the X axis), which is the U axis.

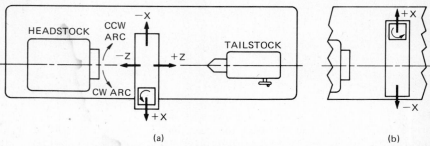

(a)   (b)

Fig. 11-3 N/C engine lathe axis notations. (*a*) With turret at front of lathe; (*b*) with turret at rear, on slant bed or vertical.

Table 11-2

Main axis	First parallel axis	Distance to arc center* parallel to main axis
X	U	I
Y	V	J
Z	W	K

* Explained later in this chapter.

In all cases the plus and minus signs are used to indicate the **direction** of incremental cutter movement (no quadrant notations are used). These directions are standardized, and even though they appear somewhat different on the lathe pictures, they are the same as those shown on Fig. 4-5.

These directions are noted in Figs. 11-3 and 11-4. In words, the notations used in this chapter are described as follows:

$+X$ = movement of cross slide **away** from center of spindle
$-X$ = movement of cross slide **toward** center of spindle
$+Z$ (or $+W$) = movement of carriage (or saddle) **away** from headstock
$-Z$ (or $-W$) = movement of carriage (or saddle) **toward** headstock, to operator's left

(It is not necessary to use the plus sign, since the MCU considers that all numbers are positive unless a minus sign is used.)

Several N/C lathes and turret lathes today mount the cross-slide turret on the far side of the workpiece. Some mount it on a "slant bed" as shown in Fig. 1-6, and some on a vertical support, as shown in Fig. 11-1. This makes a very rigid support and allows the chips to fall free. However,

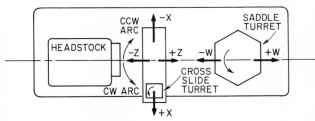

Notice that the saddle turret indexes
CCW, for the operator's safety.

Fig. 11-4   N/C turret lathe (NCTL) axis notation.

as shown in Fig. 11-3, the plus and minus *X* notation is specified so that the definitions shown above still apply. Thus the programming is not changed.

Code for Dimensions

The maximum distance which can be programmed in one block, along any axis, was, until about 1974, only 9.9999 in. Today the standard distance is often 99.9999 in. or 999.999 mm.

With the earlier limitation, a turning distance of 15 in. had to be programmed in two blocks, each usually 7.5 in. long. Each block would be programmed as *Z* − 75, as most turning cuts are toward the headstock. Many lathes now in use require this type of programming.

Our newer N/C lathe, with its longer allowable programmable distance, seldom requires "splitting" a distance.

Notice that the earlier N/C lathes used a five-digit *X* and *Y* coding. Today most lathes use six digits, with seven digits occasionally available as an optional extra.

As our coding for metric dimensioning is 999.999, the decimal point for *X*, *Y*, i, and k is understood to be after the third digit from the right. Thus 8.5 mm in *X* would be coded X0085, as trailing zeros are omitted. Leading zeros must be included.

Incremental (Delta) Dimensioning

The dimensioning of N/C programs for the point-to-point machines considered so far has been with absolute, or coordinate, measurements. This is used for the majority of the NPC machines. However, many contouring systems use the **incremental** system of dimensioning. A brief example of incremental dimensioning was given in Chap. 4. Since both old- and new-style numerical control lathes use this system, it is used for the program in this chapter.

However, most of the newer N/C lathes also have the capability of programming with absolute dimensions. The choice is then up to the programmer, who merely uses g90 and g92 for absolute, or g91 for incremental, dimensions, whichever is simpler for the particular part being programmed.

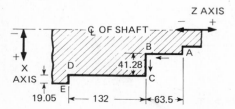

Fig. 11-5 Example of straight step turning of a shaft.

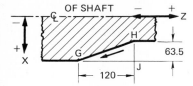

Fig. 11-6 Example of a sloped turning cut on a shaft.

   In the incremental system, the dimensions written in the program manuscript (and punched into the tape) are the actual distances which the cutter moves. The distance always begins where the cutter was at the end of the preceding command. Thus, when the cutter moves from A to B in Fig. 11-5, it moves zero mm in X and -63.5 mm in Z. When zero movement is wanted in N/C lathe programming, only the letter is written. Thus the move from A to B would be written

$XZ - 0635$

   In moving from B to C, the cutter moves only in the $+X$ direction; so this move would be programmed

$X + 04128Z$

Notice that the plus and minus signs show the **direction** of the move. The plus sign could, of course, be omitted.
   In Fig. 11-6 a cut from H to G would be programmed

$X0635Z - 12$

However, if the cut were from G to H, the signs of X and Z dimensions would be reversed, since the tool would be moving in the opposite directions. The cut from G to H would be programmed

$X - 0635Z12$

   As a check on the arithmetic and the programmer's proper use of the plus and minus signs, the program manuscript form frequently has two extra columns at the left of the page. In these columns the programmer keeps a record of the **absolute** dimensions from the zero point. The last move of the cutter is back to zero, and these check columns should also algebraically add to zero. If they do not, an error has been made. This is shown in the first two columns of Fig. 11-14.
   Numerical control lathe controls use codes which are very nearly those recommended by EIA and AIA. Each make has some differences, but these are easily learned. The codes listed in this chapter are all used by several N/C lathe companies, though no one company uses all the codes exactly as shown.

Feed Rate

On many N/C lathes made before about 1974, a special "feed rate number" (FRN) had to be calculated. One method was:

$$FRN = \frac{10F}{L}$$

where FRN = feed rate number; maximum value, 500.0 (f5), minimum value, 000.1 (f = 0001)

F = desired feed rate, in mm/min **along the tool path**

L = length of tool path, which is incremental X or Y distance if cut is parallel to an axis, or actual length of slope (hypotenuse) of a cut at an angle to the axis

An FRN of 500 actually means that the cut will take $^1/_{50}$ min (1.2 sec) to complete.

For example, in Fig. 11-5, assume that a feed of 200 mm/min is wanted. Then the FRN from A to B is

$$FRN = \frac{(10)(200)}{63.5} = 31.5 = f0315$$

FRN from D to E is

$$FRN = \frac{(10)(200)}{19.05} = 105 = f105$$

However, in Fig. 11-6, the path of the cutter is sloped, so L in the formula is the hypotenuse of the right triangle GHJ. If the drawing is accurately made and not too small, it is sufficiently accurate to scale the distance HG. It can be computed by the Pythagorean formula, which is

$$GH^2 = HJ^2 + GJ^2 \quad \text{(Appendix A)}$$

Thus

$$GH^2 = 63.5^2 + 120^2 = 4032 + 14\ 400 = 18\ 432$$

Therefore

$$GH = \sqrt{18\ 432} = 135.8 \text{ mm}$$

This figuring is sufficiently accurate if done with a slide rule; or the tables of squares and square roots in a handbook may be used. Thus the FRN for the cut from H to G (or G to H) is

$$FRN = \frac{(10)(200)}{135.8} = 14.7 = f0147$$

When cutting a **circular path,** the formula for feed rate number is very similar; for circular arcs,

$$FRN = \frac{10F}{R}$$ using circular interpolation

where F = desired feed rate, ipm along the circular arc
R = radius of arc, in.

There are many of these older lathes still in use, and computing the FRN is not difficult. However, the computations vary, so it is necessary to check the programming manuals.

Table 11-3 CODING AND SPECIFICATIONS FOR THE N/C ENGINE LATHE USED IN CHAPTER 11

1. Word address program, use of tab permitted but not required.
2. Omit trailing zeros. Leading zeros must be used.
3. Incremental (delta) dimensioning. Absolute dimensioning available with g90.
4. Linear and circular interpolation is standard equipment.
5. Maximum $X, Z, i,$ or k is 99.9999 in. or 999.999 mm.

Letter codes		Preparatory functions		Miscellaneous functions	
f	Feed rate mm/min.	g01	Straight-line cut (axial or slope)	m00	program stop
				m01	optional stop
n	Sequence number	g02	CW arc	m02	End of program†
		g03	CCW arc	m03	Spindle on-CW rotation (reverse)
i	X offset (or sine)	g04	Dwell (use X register for time)*		
				m04	Spindle on-CCW rotation (forward)
k	Y offset (or cosine)	g33	Thread, constant lead		
		g90	Absolute dimensions	m05	Spindle stop
s	Spindle speed (coded table)	g91	Incremental dimensions	m08	Coolant on
		g94	mm/min feed rate	m09	Coolant off
		g95	mm/rev feed rate	m30	End of tape
t	Turret positions (eight available)				
X	Cross-slide motion				
Z	Carriage motion				

Note: $X, Z,$ i, k, m00, m01, m02, m30 are not modal; they are canceled at the end of each block. All other commands remain in memory until changed or canceled by m02 or m30.

* Use a 1-sec dwell with all tool changes, and a 3- to 10-sec dwell for speed changes unless they are relatively small and in the same range.
† Use of m02 or m30 varies with the MCU being used.

Today, practically all the new N/C lathes use feeds specified directly in millimetres per minute or inches per minute, with sometimes optional mm/rev or ipr feed rate programming using g94 and g95 (Table 11-3). The mm/min is used in this chapter. The range of available feeds will vary with the model and manufacturer. Our lathe has feeds from 0.1 to 5000 mm/min (f00001 to f5).

Spindle Speeds (S)

The S code letter is always used before the spindle-speed code number. Many N/C lathes use a two- or three-number code. This code number usually refers to a table of the speeds available on the particular make and model of lathe being used. In a three-number system such as that shown in Table 11-4, the first digit refers to a speed range, and the next two digits specify selection within this range. From two to five speed ranges may be available. In our example there are three speed ranges, with 13 speeds in each range, a total of 39 speeds. This table does not represent the actual speeds of any particular lathe. It is used in this chapter to illustrate the type of coding that may be used.

Notice that the rpm's in speed ranges 1 and 2 each interlock with the speeds in the next-higher range. This is frequently, though not always, the

Table 11-4   CODING FOR SPINDLE SPEEDS AVAILABLE ON THE N/C ENGINE LATHE USED IN CHAP. 11

Code	Range 1, rpm	Code	Range 2, rpm	Code	Range 3, rpm
100	15	200	60	300	240
101	17.5	201	70	301	280
102	20.4	202	82	302	328
103	23.8	203	95	303	380
104	27.7	204	111	304	444
105	32.2	205	129	305	516
106	37.5	206	150	306	600
107	43.6	207	174	307	696
108	50.7	208	203	308	812
109	59.0	209	236	309	944
110	68.6	210	275	310	1100
111	80.0	211	320	311	1280
112	92.0	212	368	312	1472

case with commercial N/C lathes. The maximum and minimum speeds available depend considerably on the size of the lathe. A lathe with a 460-mm swing over the bed can machine only relatively light work; so the top speed might go up to 2000 rpm. However, if a lathe has a 1520-mm swing over the bed, it might be efficiently used at as low as 3 rpm, and possibly at no faster than 750 rpm, since it will be used mostly for large, heavy work.

On our lathe, if the computed speed were 278 rpm, a speed could be used from range 2 in Table 11-4 (275 rpm, coded S210) or from range 3 (280 rpm, coded S301). It is usually better to use speeds near the top of the speed range when there is a choice; thus speed S210 would probably be used.

Dwell Time (g04)

When programming a change in the speed, or a turret index, it is often necessary to program a dwell. This stops all tape-reader and X and Z movement.

If the change in speed is large, a 2- to 10-sec dwell may be required. If the speed change is only one or two steps within the same range, the dwell may, on some lathes, be omitted.

When indexing the turret, a 1- or 2-sec dwell is required to get the motion started. After this time, an interlock is activated which prevents any X, Z, or tape motion until the turret is in position.

The DWELL code is g04 (Table 11-3), and the number of seconds of dwell is specified as an X command, with a maximum of 99.999 sec. Thus a 2-sec dwell is programmed as

g04 X02   trailing zeros may be omitted

Notice that the length of the dwell is programmed as an X function. The g04 tells the MCU that this is a dwell; so the MCU sends this signal to the proper control section. The X function is frequently, but not always, the one used for dwell times.

Do not program any X or Z commands in the same line with a g04. A short dwell may be specified at the bottom of a necking cut, or when boring or turning to a shoulder, but this is not required.

Turret Position (t)

The turret referred to in this chapter is a turret on the **cross slide,** not on the saddle, as in a conventional turret lathe. Many engine lathes use a square turret on the cross slide even in conventional machining, because this makes four or more tool bits available without changing the setup. The four-sided cross-slide turret is also often used on N/C lathes. Other turrets available are six-sided turrets, round turrets, and sometimes a multiple rear toolholder on the cross slide.

The front cross-slide turret is indexed by tape control to one of a number of positions. Two methods of assigning these positions are shown in Fig. 11-7.

The t command on engine lathes has either two or three digits. In our machine we use three digits. The first digit calls for one of the eight turret

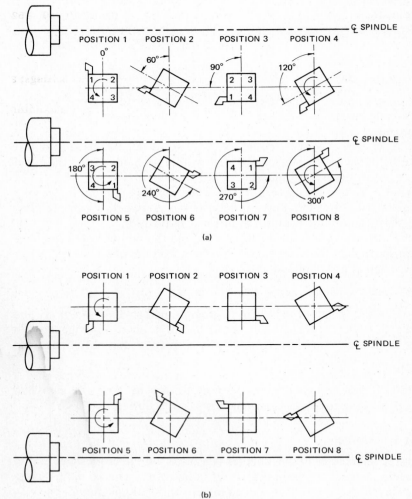

Fig. 11-7 Indexing positions of square cross-slide turret. (*a*) With turret at the front; (*b*) with turret at rear, often on a slant bed.

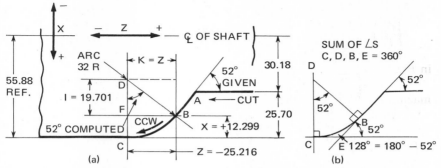

Fig. 11-8 Example of a convex arc cut as it might be turned on a shaft.

positions. The next digits call for one of 32 available **tool-offset settings;** a zero in these digits would mean no tool offset.

The tool offset is dialed in at the control panel and has a maximum value of ±9.999 mm on our lathe. Thus, if the tool bit wears a few hundredths of a millimetre, or if the original setup is slightly off, the adjustment can be made by the dial-in method. Any one of the tool offsets may be used with any one of the turret positions. For example, all cutting tools could use the same offset.

Typical turret coding would be t302, which would mean turret position number 3, with the offset which has been dialed into position-2 offset control station.

Arc Offset Commands (i and k)

Most N/C engine lathes are supplied with linear-interpolation capabilities. That is, they can cut straight-line slopes at any angle, as illustrated in Fig. 11-6. A programmer can compute straight-line cuts to approximate a circle closely, but this can be a long job. So almost all N/C lathes are supplied with added controls for circular-interpolation capability (g02 and g03).

Circular interpolation means that the machine will, with only one block of information, drive the cutter along the path required to machine an arc of a circle. On our N/C lathe, the arc cannot be more than 90° in one block, and it must be entirely in one quadrant. If the arc is over 90°, or crosses into another quadrant, another block of information must be coded. The radius of the arc must not be over 999.999 mm.

A typical arc is shown as *BC* in Fig. 11-8*a*, and the block of information needed includes the following information:

g02 or g03 CW or CCW arc. This direction is indicated by thinking of the plan view of the workpiece (looking up from below) as it is secured in the lathe. The arc *BC* in Fig. 11-8 is counterclockwise if the tool moves from *B* to *C*, along the outside of the workpiece (Figs. 11-3 and 11-4).

Note: CW and CCW arcs described are according to EIA. However, some N/C lathe manufacturers use the opposite convention; so a programmer should check before using g02 and g03 on each N/C lathe.

X **distance** actually moved, with plus or minus indicating the direction.

Z **distance** actually moved, with plus or minus indicating the direction.

I **(offset)**   is the distance along the *X* axis from the **starting** point of the arc to the center of the arc. The offset may be from 0.0000 to 99.9999 mm. No plus or minus sign is needed.

K **(offset)**   is the same as *I*, but measured along the *Z* axis, from the start of the arc to the centerline of the arc. No plus or minus sign is needed.

f, s, and t   are the same as in any block of information.

In order to get the *X, Z, I,* and *K* dimensions, someone must do some mathematics. The computations might be done by the designer or drafter, or the part programmer may have this responsibility, or a computer program might be used. We assume that the programmer must do the mathematics for the dimensions of Fig. 11-8.

Figure 11-8*b* shows how, with a little geometry, you can prove that angle *CDB* in quadrilateral *BECD* equals 52°. In a five-place trig table in a reference book we find that

$$\sin 52° = .78801$$
$$\cos 52° = .61566$$

In triangle *FDB*,

$$\frac{DF}{DB} = \cos 52° \qquad \text{therefore} \qquad DF = DB (\cos 52°)$$

$$DF = I = (32)(.61566) = 19.701$$

Also

$$\frac{BF}{DB} = \sin 52° \qquad \text{therefore} \qquad BF = DB (\sin 52°)$$

$$BF = Z \qquad \text{and also} \qquad BF = K = (32)(0.78801) = 25.216$$
$$X = DC - DF = 32 - 19.701 = 12.299 \text{ mm}$$

With these figures now available, the program for cutting arc *BC* in Fig. 11-8*a* would be written

g03 *X*012299 *Z* − 025216 *I*019701 *K*025216 (plus feed and speed codes)

Figure 11-8 shows how this cut would look if it represented a contour

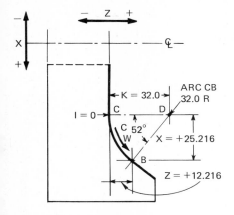

Fig. 11-9 Example of a concave arc cut as it might be faced on a casting.

on the outside of a shaft or similar bar-stock job. Figure 11-9 shows the same contour on a chucking job, cutting from the center outward. Even though this reverses the figures on the X and Z axes, the computations are the same, except that, because the cut starts in line with the arc's center, I is zero. The cut is now in a clockwise rotation.

The program for cutting the arc CB in Fig. 11-9 would be written

g02 X025216 Z012291 I K032 (I with no digits indicates a zero coordinate)

Nose Radius of Tool Bit

If there were a sharp point on all the cutting tools used on a lathe, the Z distances programmed for a cut such as shown on Fig. 11-10 would be Q, R, and S, as shown. However, tool bits usually have a nose radius, and as shown on Fig. 11-10, this changes the point of contact between the tool and the workpiece, and can make a difference of several hundredths of a millimetre in the dimensions actually cut, compared with the dimensions desired.

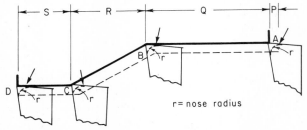

Fig. 11-10 Comparison of the cutting paths of tool bits with sharp points and tool bits with nose radius.

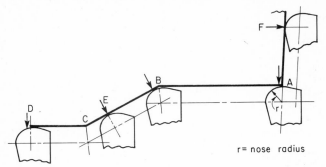

Fig. 11-11   Enlargement of the effect of nose radius on cutting path in turning slopes.

The necessary correction may be made easily at point *A*, where the distance *P* is equal to the nose radius *r*. However, at points of intersection between slopes and straight cuts, as at *B* and *C*, the correction is not as simple.

To see more exactly what happens, refer to Fig. 11-11. This shows an enlarged view of the effect of nose radius on the actual cutting path. At points *A* and *D*, the front edge of the tool bit is doing the cutting. At point *F*, the side of the tool is cutting, and at point *E*, the cutting is being done at some point in between the front and side. Of course, there will be a radius at point *B*, which the machine designer must consider.

The solution to this problem may be visualized by considering the nose radius as a small end mill. When cutting an edge with an end mill, you program all dimensions to the center of the cutter. The same method may be used for lathe cuts; that is, program the *X* and *Z* dimensions to the center point of the nose radius. This path is shown by the dotted lines in Figs. 11-10 and 11-11. The formulas for computing the position of this center are not complicated. However, in order to concentrate on the method of programming, we consider that we are using a tool bit with a sharp point and omit any consideration of the effect of the nose radius. Incidentally, computer programs are available which will figure the cutter offset path and give the corrected *X* and *Z* dimensions, plus all other needed data.

An N/C Lathe Program

A cross section of the part to be machined is shown in Fig. 11-12. This is a flange which has already been machined on the opposite side, on the same lathe, including the 50.0-mm center hole. The job is to face the entire front and shaped surfaces (points 1–2–3–4) and turn the outside edge to 350.0-mm diameter, to point 5. The method used is not the quickest (the

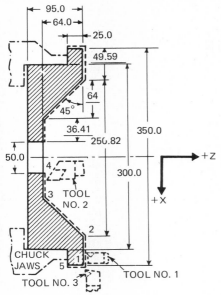

FLANGE

Material — mild C.I.
Approx. 3 mm stock to be removed.
Part previously machined on other face.
3 jaw chuck, jaws turned and faced to
fit work piece.
Cutting speed = 60 m/min
Feed = 0.30 mm/rev

Fig. 11-12 Drawing of workpiece, to be machined as shown. The cutting tools are shown adjacent to the surfaces on which they are used. See Fig. 11-14 for the N/C program.

facing tool in position 3 of Fig. 11-13 might be used for all facing), but is designed to illustrate several ideas which must be considered in programming an N/C lathe.

The tooling to be used is shown in Fig. 11-13. All tools are 25-mm-square toolholders with carbide inserts. The 60-m/min cutting speed is conservative, and the 3.0-mm depth with 0.30 feed can be easily handled.

The order of operations is as follows:

1. Face (1 to 2) with tool 1.
2. Face the flat (4 to 3) with tool 2.
3. Face the slope (3 to 2) with tool 2.
4. Turn 350 mm O.D. (1 to 5) with tool 3.

Setup and Zero

The cutting tools would be positioned in the 178-mm [7-in.] square turret as shown, with the edges of the tools lined up with the sides of the turret. If these cutters had a nose radius, special equipment is available which would make it a relatively simple job to position the center of the nose radius at the desired dimension. In either case, the turret may be set up while not on the machine.

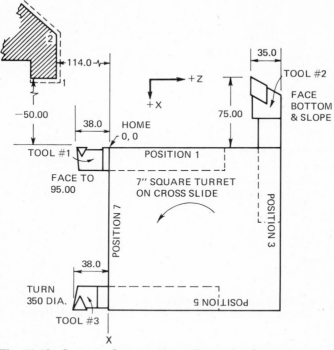

Fig. 11-13   Layout of square turret (located at front of cross-slide), showing setting dimensions for cutting tools, as used in this chapter.

Numerical control lathes may have dimensions marked on the side of the bed and on the side of the cross slide. These dimensions are usually measured in the X direction from the centerline of the spindle and in the Z dimension from the face of the spindle, or from a known distance out from the face.

By using these markings plus the manual-input dials on the console, or resolver dials, the turret location can be set within a few thousandths of its final position. The tool-offset dials can then be used to complete the settings during one or two trial cuts. Other methods are also used, but they will not be given here because the operator learns them on the job.

The N/C Engine Lathe Program

The complete program for machining the face and edge of Fig. 11-12 is shown in Fig. 11-14. An additional sketch, Fig. 11-15, is shown, with the **absolute** X and Z dimensions from the zero point. These dimensions would probably be marked on the part print which the programmer is using. The **underlined** dimensions in Fig. 11-15 were added to help in figuring the incremental distances used in this program.

Explanation of Program of Fig. 11-14

Set up (see Fig. 11-13) at the corner of the tool-post, with $X = 175.0 + 50.0 = 225.0$ mm from the centerline of the spindle, and $Z = 114.0$ from the finished face of the workpiece (which is 25.0 mm from the face of the chuck). This will be 0.0, 0.0 for our incremental programming dimensions.

n001   Start-up.

g04   Commands a DWELL.

X02   Calls for a 2-sec DWELL to allow the turret to start indexing.

s108   rpm $= \dfrac{300(60)}{350} = 51.4$, use 50.7, s108.

t101   Turret position 1, with tool-offset setting 1.

m03   Commanding CW (forward rotation) of the spindle.

Note: g91, incremental dimensioning, is the normal mode, so it does not need to be programmed.

n002   g94   Specifies mm/min feed rate programming.

ABSOLUTE DISTANCE FROM START		SEQ. NO.	PREP FUNCT.	INCREMENTAL DISTANCE		DISTANCE TO ARC CENTER		FEED RATE	SPIN. SPEED	TOOL NO.	MISC. FUNCT.	REMARKS
				CROSS AXIS	LENGTH AXIS	PARALLEL TO X	PARALLEL TO Z					
X AXIS	Z AXIS	N	G	X±	Z±	I	K	F	S	T	M	
		NOTE :	This program sheet to be used for METRIC DIMENSIONS ONLY									
0	0	n001	g04	x02					s108	t101	m03	g91, NORMAL MODE
		n002	g94									mm/min FEEDS
-45.50	-76.00	n003	g01	x-0455	z-076			f5			m08	RAPID APPROACH
-105.09		n004	g01	x-05959				f00152				FACE END (1 TO 2)
	0	n005			z076			f5				RETRACT TURRET
		n006	g04	x02					s110	t302		CHANGE "S" & INDEX
-237.50	-90.00	n007	g01	x-13241	z-09			f5				RAPID APPROACH
	-103.00	n008			z-013			f00215				" "
-198.59		n009		x03891								FACE FLAT (4 TO 3)
-132.09	-36.50	n010		x0665	z0665			f00137	s110			FACE SLOPE (3 TO 2)
	0	n011			z0365			f5				RETRACT TURRET
		n012	g04	x02					s108	t703		CHANGE "S" & INDEX
-12.0	-111.50	n013	g01	x12009	z-1115			f5				RAPID APPROACH
	-141.50	n014			z-03			f00152				TURN OD (1 TO 5)
0	0	n015		x012	z1415			f5			m05	TRAVERSE HOME
		n016	g04	x02						t10	m09	1ST TOOL, COOLANT OFF
		n017	g								m30	END, REWIND TAPE

At the top of the sheet:

N/C ENGINE LATHE PROGRAM SHEET
MATERIAL _CAST IRON_
CUSTOMER _____
PART NO. _FIG. 11-15 & 11-16_
PART NAME _FLANGE_
OPER. NO. _____
TAPE NO. _____
PAGE _1_ OF _1_
DATE _____
PROGRAMMER _____

Fig. 11-14   The metric N/C program manuscript for turning and facing the workpiece shown in Fig. 11-12, on the N/C lathe specified in this chapter.

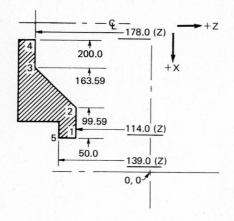

Fig. 11-15 Absolute X and Z locations to machining locations.

n003 Rapid approach to point 1, facing.

    g01   Specifies a straight-line movement.

    X – 0455   Move in toward the center

– 50.00	tool point to finished O.D.
+ 3.00	allow for rough surface
– 47.00	
+ 1.50	clearance
– 45.50	mm to move in

    Z – 076   Move toward the spindle for facing

– 114.0	from face of turret to face of part
+ 38.0	projection of the tool from turret face
– 76.0	mm

    f5   5000 mm/min—rapid traverse

    m08   Turns on coolant

n004 Face the end, positions 1 to 2.

    g01   Repeated.

    X – 05959   Cross slide moves inward to face 1 to 2.

– 49.59	width of flange face (Fig. 11-12)
– 3.00	rough surface outside
– 3.00	rough surface inside
– 1.50	previous clearance
– 2.50	clearance beyond point 2
– 59.59	mm

    f0152   50.7 rpm $\times$ 0.3 = 15.2 mm/min

n005 Retract turret (in Z direction only).

Z + 076   Moves turret + 76 mm *away* from the spindle, back to Z zero position.

f5   Rapid traverse.

Note: It is necessary to move the turret back this far so that when it indexes, in a CCW direction, the cutter will not bump into the workpiece. This is important to watch for, not only on lathes but in all numerical control equipment.

n006 Change speed and index turret.

g04 and X02   Give a 2-sec delay for start of turret indexing.

s110   Faster, as part diameter is smaller.

$$\text{rpm} = \frac{(300)(60)}{250.82} = 71.8, \text{ use s110.}$$

No DWELL needed, as speed change is small.

t302   Rotates the turret to position 3, using tool 2, and tool offset setting 2.

n007 Rapid approach to point 4 plus 2.50 clearance.

g01   Straight-line movement

X − 13241 Moves the turret 132.41 mm toward the centerline to a distance 2.5 mm beyond the edge of the 50-mm bored hole.

− 200.00	absolute distance to edge of the hole
− 2.50	clearance beyond the hole
− 202.50	
+ 105.09	present absolute *X*, from previous two moves (Fig. 11-16)
− 97.41	
− 35.00	width of tool (Fig. 11-16)
− 132.41	mm to point *B* (Fig. 11-16)

Z − 09   Moves the turret to within 13 mm of the face of surface 4-3.

− 114.00	edge of turret to face of flange (Fig. 11-16)
+ 75.00	projection of tool (subtracting a negative)
− 39.00	tool point to face of flange
− 64.00	flange face to surface 3-4
− 103.00	total move to finished edge
+ 13.00	stay back to avoid collision
− 90.00	mm actual movement in Z

f5   Rapid traverse.

Note: Figure 11-16 shows the direction of motion of this cut. It would be desirable to program a move directly from *A* to *B* in this figure. However, it was found that this path would cause the point of the cutter to bump into the 3.0-mm rough surface just before point 4. This possibility showed up when the cutter path was drawn to scale on the original print. A little mathematics proved that a collision *would* occur.

To avoid this collision, the 2.50-mm clearance from point 4 to the tool point could be increased to a minimum of 4.10 and just avoid the corner. Or the turret could be moved back to start at a zero point, 172 mm (instead of 114 mm) from the part; but this makes a longer path. Thus programming to a point 13 mm away from point 4 in line 007 and then programming straight forward for 13 mm in line 008 seems a quick, safe solution to the problem.

n008

Z − 013   To move toward the spindle, in line with surface 3-4.

n009

X + 03891   To face surface 3-4.

$$+\ 36.41 \qquad \text{length of face 3-4 (Fig. 11-12)}$$
$$\underline{+\ \ 2.50} \qquad \text{clearance (used in n007)}$$
$$+\ 38.91 \text{ mm move}$$

f02154

Feed rate = 0.3 × 71.8 = 21.54 mm/min.

n010 Cut 45° surface.

*X*0665, *Z*0665   Both axes are the same as the tool moves on a 45° slope, in a plus direction, along path 3-2.

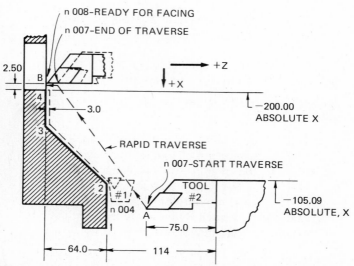

Fig. 11-16 Tool path to position cutting tool for facing cut (4 to 3 to 2) to avoid hitting the work at point 4.

+ 64.00   X and Y coordinates of the 45° slope
+ 2.50   clearance beyond point 2
+ 66.50 mm

s110

$$\text{rpm} = \frac{(300)(60)}{250.82} = 71.8, \text{ use s110, 68.6 rpm}$$

f0137   Use only 0.2 mm/rev feed, as the tool is cutting on a 45° slope. Thus

$$68.6 \times 0.2 = 13.72 \text{ mm/min}$$

n011   Retract turret to Z = zero to allow for tool change.
Z + 0365

− 114.00 absolute position of finished surface
+   2.5   added clearance in n010
− 111.50 present position of tool point
+ 75.00 projection of tool from turret
− 36.50 distance away from absolute zero.

Thus a + 36.50 will return the turret to zero.
X   No change necessary, though a move could be made.
f5   Rapid traverse
n012   Change tool and speed.
g04 and g02   DWELL for tool change.
s108   Same speed as in n001, as same diameter.
t703   Index to turret position 7, using preset offset 3, brings tool 3 in position for facing 1–5.

n013
g01   Repeated, as g04 canceled it. The turret is now in the position shown in Fig. 11-17. It must be moved to start facing at point 1.

X + 12009

− 99.59 absolute distance to point 2 (Fig. 11-15)
+   2.50 clearance in n010
− 97.09 last X location of point of tool
− 35.00 width of toolholder
− 132.09 present X of edge of turret (Fig. 11-17)
− 38.00 to cutting edge of tool 3
− 170.09 present X of tool point
+ 50.00 to finished dimension of O.D.
− 120.09 X distance to move, but sign will be plus, as move is away from the machine centerline.

Z − 1115

> − 114.00 to move point to work location 1 (Fig. 11-16)
> + <u>   2.50</u> clearance
> −111.50 total move for Z

     f5   Rapid traverse.
n014 Z − 03

> − 25.00   thickness of flange
> − 2.50   clearance at start of cut
> − <u> 2.50</u>   clearance at end of cut
> − 30.00 mm total move to turn 1–5

    f0152   Same as in n004.
n015 Move both axes, in the plus direction, back to the original 0.0, 0.0 point (home position).
    X012

> + 50.00 edge of work to X = 0.0
> <u>− 38.00</u> tool projection (Fig. 11-13)
> + 12.00 to return X to zero

    Z + 1415

> − 114.00   face of part to Z = 0.0
> − 25.00   width of part
> − <u> 2.50</u>   clearance used in n014
> − 141.50 mm present Z absolute position, so + 141.50
>   will move the turret to home position

    f5   Rapid traverse.
    m05   Stop spindle.
n016
    g04   X02 DWELL for turret indexing.
    t10   Turret position 1, and cancel any offset.
    m09   Coolant turned off.
n017 m30 (or m02)   End of program, rewind the tape.

Turning and Threading Shafts

Turning shafts, long or short, and similar shapes, between centers follows the same procedures as in the above example. In fact the distances are often easier to compute, and it is not always necessary to have a tool change.

    Grooving cuts can easily be made by using the proper-width toolbit, positioning in Z to the proper location, then programming a minus X to depth, than a plus X to back out.

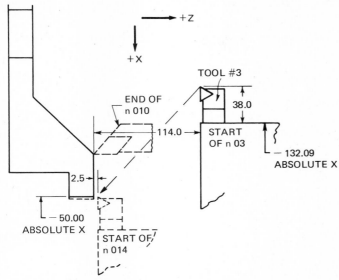

Fig. 11-17   Tool path to the beginning of the turning cut on the 350-mm O.D.

Threading is most often done with g33 code and the k column, for longitudinal threads. The number following the k is the lead (advance of the thread in each revolution). In the English system the lead is 1/threads per inch. In the metric system the lead, in millimetres, is part of the thread specification.

For example, a $^1/_4$-20 UNC thread has a lead of $^1/_{20}$ = 0.050 in. so k0005 would be programmed. The metric close equivalent is specified M6 × 1.0 [6 mm dia. with a 1.00 mm lead], so k001 would be programmed (see Table 11-1).

Conclusion

Slopes, curved surfaces, and chamfers are easily programmed on lathes, thus eliminating the need for templates. The lathe manufacturers have many special programs and other aids which simplify the programmer's work.

The next chapter shows N/C applied to a very different type of machine. The use of N/C has often more than doubled the capacity of these ''hole punchers,'' as it has for lathes also.

Chapter **12**

PROGRAMMING OF N/C HOLE-PUNCHING MACHINES

The machine studied in this chapter, unlike those previously studied, does not have a rotating spindle, or a rotating work-holding device. This machine is called a hole-punching machine, or a fabricator, or a turret punch press, or just an N/C punch press.

These machines have been used for several years for punching hole patterns in electrical chassis and cabinets, instrument panels, splice plates, machine cabinets, thousands of small gratings, etc. Combining numerical control with them has, according to several reports, produced three to fifteen times as much work, with better accuracy and fewer rejects. Thus today these N/C machines outsell the less expensive old-style hole punches which used layout and templates to create the patterns.

Today these machines punch a wide variety of shapes and sizes. Some manufacturers have added tapping attachments, and one has a plasma-arc torch cutting attachment for cutting large or irregularly shaped openings—and all under the same N/C control.

While the most widely used fabricators are of 267- or 356-kN [30- or 40-ton] capacity, some are made to over 890-kN [100-ton] size and are able to pierce 28-mm [1⅛-in.] thick mild steel plate.

The smaller presses can sometimes make as many as 200 "hits," or strokes, per minute, though most work is done at rates of 60 to 120 hits per minute.

This, as you will see, is one of the simpler machines to program. However, there are often a large number of holes, or repeated hole patterns, so programming without a computer can take considerable time.

Each of the four major companies which make N/C punching ma-

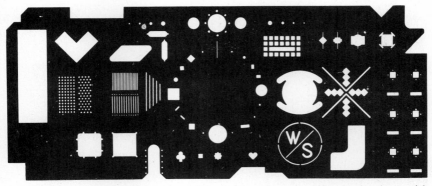

Fig. 12-1 A sheet showing some of the types of work which can be done with a hole-punching machine. Center group shows one set of standard punches in the turret. Large holes were "nibbled" out. (*Wiedemann Division, Warner & Swasey Co.*)

chines uses a different coding system. However, they stay fairly close to the EIA suggestions, so it is not difficult to learn each system. The coding used in this chapter is typical of one major company, though to keep this a basic text, the complete system is not used.

Work Done

(See Fig. 12-1)

Punching holes—round, square, hexagonal, or any special shape, from about 1- to 125-mm diameter.

Notching—usually, but not always, square notches at corners or edges.

Nibbling—making a series of short steps, to form a curved, or diagonal cutout. The edges will be slightly scalloped, so they may need to be smoothed.

Knockouts, for electrical connections, are pierced, except for one or two small tabs which hold the slug in place.

Louvres, for ventilation, are pierced and formed in one stroke.

Countersinking, extrusions, corner radii, and **keyhole shapes** are also done.

The Punches

The punch and the matching die are made in the shapes shown in Fig. 12-2. There are round punches from about 2 to 150 mm in diameter, square punches up to 50 mm square, etc., in different punch-die clearances for different materials and stock thicknesses.

These die sets are made to fit different sizes of holders in the turret.

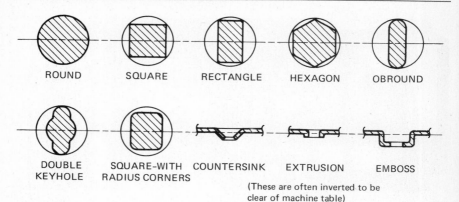

ROUND   SQUARE   RECTANGLE   HEXAGON   OBROUND

DOUBLE   SQUARE—WITH   COUNTERSINK   EXTRUSION   EMBOSS
KEYHOLE   RADIUS CORNERS

(These are often inverted to be
clear of machine table)

Fig. 12-2   Some of the more frequently used shapes of punches.

The Turret

N/C hole-punching machines are made with a single punch holder or with multiple holders arranged in a circular turret, much like some of the tool-holders shown in Chap. 10.

The **single-punch** machines can handle larger sheets of metal and are made with greater tonnage than the turret types. The M06 tool change code is usually used, and a punch and die set can often be changed in less than a minute. As example of this type of machine is shown in Fig. 12-3. This has the optional plasma-arc cutter. A tapping attachment is also available.

The **turret-type** machines may have from 10 to 38 die holder stations. The machine we will use has 20 stations. Because it would not be economical to make all stations heavy enough to punch large-diameter holes, the manufacturers provide a few stations for the heavy, larger punch and die sets. Our N/C punch press uses heavy, 75-mm [3-in.] stations at positions 1, 6, 11, and 16. The other sixteen stations hold tooling up to 32-mm [1¼-in.] in size. This type of turret punch is shown in Fig. 12-4.

Size of Work Done

Many parts which are made on N/C hole-makers are only 600 to 900 mm [2 to 3 ft] long, and often only half this in width. These easily fit on even the smallest punching machines.

However, much larger sheets of metal are often required. All punching machines have a "nominal" or standard-size sheet capacity listed. This may be from 760 × 910 mm [30 × 36 in.] for a 267-kN [30-ton] punch to 1210 × 1520 mm [48 × 60 in.] for a 534-kN or 1.02-MN [60- or 115-ton] heavy-duty fabricator.

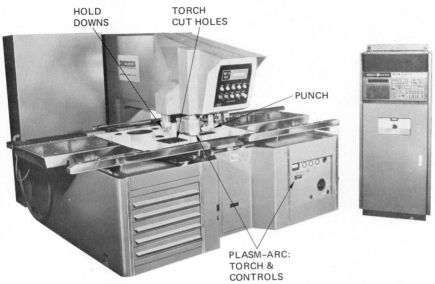

Fig. 12-3   A hole-punching machine, single punch, quick change tooling. Shown with optional plasma-arc torch for cutting large openings faster than the nibbling. Capacities up to 1.02 MN [115 tons]. Sheet size up to 1200 mm [48 in.] wide, or double if sheet is turned 180°, by any length. (*W. A. Whitney Corp., an Esterline Company.*)

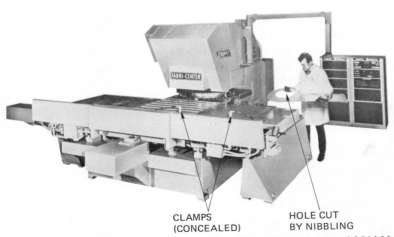

Fig. 12-4   A turret hole-punching machine. Up to 38 stations and 356 kN [40 tons] capacity. Maximum throat depth 1295 mm [51 in.]. Any length can be handled by using repositioning clamps. (*Strippit Division, Houdaille Industries, Inc.*)

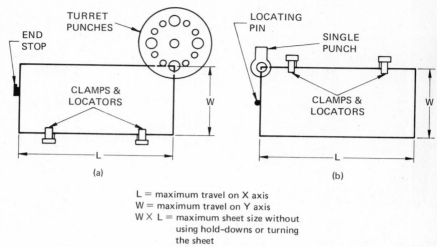

L = maximum travel on X axis
W = maximum travel on Y axis
W X L = maximum sheet size without
using hold-downs or turning
the sheet

Fig. 12-5 Illustrating maximum sheet size and the positions of the clamps in the two types of fabricating machines.

The lengths (see Fig. 12-5) can, in many machines, be much greater, as special temporary hold-downs make it possible to relocate the work clamps one or more times.

The widths, in the turret-type machines, cannot be increased, as the throat distance, from work clamps to the centerline of the punch, is fixed (Fig. 12-5a). However, since the work clamps in a single-station hole puncher can pass behind the punch, they can be located on the punching side of the table. Thus, a wide sheet can be turned 180°, effectively doubling the allowable width (Fig. 12-5b). Of course, additional work-tables must be used to hold these larger plates in either direction.

The same as in all N/C or standard machines, the actual table movement in X and Y is built in and cannot be changed.

Materials Punched

The greatest bulk of hole punching is done on mild steel sheet and plate with a shear strength of 345 MPa [50 000 psi] or less. However, galvanized steel, stainless steel, brass, fiber glass, and other materials can be handled. Of course, not all materials can be punched through the same thickness.

Material thickness is most frequently from 18 gage [0.048 in. or 1.2 mm] to 6 mm [¼ in.], though much heavier plate can, as mentioned previously be punched.

The actual maximum-size hole which can be punched depends on the shear strength of the material and the material's thickness. A typical

Table 12-1 CAPACITY OF A 356-kN HOLE PUNCH*

Punch diameter		Will punch through thickness†	
Inches	Millimetres	Inches	Millimetres
1	25.0	½	12
2½	62.5	¼	6
3½	87.5	6 ga (0.194)	5
4½	112.5	8 ga (0.164)	4
5	125.0	10 ga. (0.135)	3.5

* In mild steel, 345 MPa shear strength.
† Turret press capacity is somewhat less.

capacity table for a 356-kN [40-ton] press is shown in Table 12-1, for a single-station machine.

Work Holding and Locating

The workpieces (sheets of metal) are held by two special clamps. The clamps are often operated by small air cylinders which, in a grip of about 10 mm deep × 30 mm wide, securely hold the work and move it in the X and Y directions. These clamps, as mentioned previously, may be at either the front or back of the machine table.

Stops are built into the clamps, so that they locate the workpiece parallel to the X axis. The stops also, when at home position, locate the $Y = 0.0$.

At one end of the worktable is a side stop which locates the $X = 0.0$. This stop cannot be moved horizontally, but can be moved back and forth, to contact the end of the work at a convenient location.

There is a "no punch" area around all clamps. This is to prevent a collision between the punch holder and the clamps. This area is specified with each machine. The no punch areas for typical machines are shown in Fig. 12-6. Most machines have safety devices which will stop the entire machine if the punches get too close to the clamps. However, a good programmer specifies clamp locations at setup which will avoid any interference.

Dimensioning the Drawing

Some N/C hole-punching machines normally use absolute dimensioning, and some are set for incremental dimensions when they are started. However, practically all the machine control units made from 1976 on have G or M codes which allow either type of programming (or both in the same tape, if it is easier) to be specified.

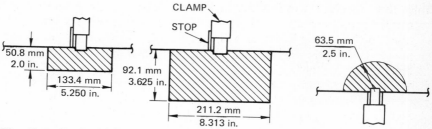

Fig. 12-6   Typical "no punch" zones around the work clamps.

As discussed in previous chapters, baseline dimensioning on the drawing greatly simplifies the programmer's work if absolute dimensions are to be punched in the tape. Our machine normally uses incremental programming, so the part to be programmed is dimensioned to make this easier.

Coding to be Used

Our N/C fabricator uses both G and M codes, as in previous chapters, and T codes for random tool selection. The programming form shown is a simplified one, as this same machine can do circular and linear interpolation (using I, J, and F codes) and continuous path nibbling. These can easily be learned and are not included here, in order to simplify the program.

The coding used is shown in Table 12-2. Most of the terminology should be clear, as it is similar to that used in previous chapters, even if the code numbers are not the same. Explanations of the following new codes will help.

G01, G02, G03   are not used in this program. They are used for linear and circular interpolation.

G67   Punch off. Stops all punching action until a G68, Punch on, code is programmed. Used for safety of operator when starting up, or in manual mode.

G69   Retract command, moves the work back to home position and turns off the punch. Next block is then read and executed. Use G68 to start punch.

M06   is used only at the beginning of this program. This rotates the turret so that tool 1 is at the punching station. This ensures that the turret will index to the proper position when T codes are used.

M12   Cycle stop. Stops the machine after completion of the work called for in that block. Push *cycle start* to continue.

M71   Tape rewind. If tape is short, it is made into a continuous loop. After the last block, tape goes to block 001 where it finds a G69 (or a G67) and stops. If tape is long, and on spools, the M71 rewinds the tape.

M75   Load position. Stops punch and tape reader. The first *X* and *Y* commands often position the clamps at a location other than home position, for loading. Sometimes this is to place the work closer to the punch, to save time in the first move. Push *start* and use G68 to start punch after loading.

T01–T20   Call for turret positions. They can be in any order. The tool station is specified in the first block in which the tool is used. The *X* and *Y* moves in this block are delayed until the turret is in position. Then the hole is punched, and the next block is read.

Programming Limits

The sequence number is N with three digits, and the G and M and T codes use two digits. Codes *X*, *Y* (and I and J when used) are:

inches ± 999.999 *or* 99.999
millimetres 9999.99
trailing zeros may be omitted

Accuracy of location is ± 0.13 mm [± 0.005 in.] or better. Repeatability tends to be better than this.

Table 12-2 CODING USED IN THIS CHAPTER

G01*	Linear nibbling mode.
G02*	Clockwise circular nibbling mode.
G03*	CCW circular nibbling mode.
G67	Punch off.
G68	Punch on.
G69	Retract to home position. Punch turned off, read next block.
G90, 92, 95*	For absolute programming mode.
G94	Incremental programming mode. As this is the usual mode when the machine is turned on, it does not have to be programmed.
M06	Tool 1 at punching station (start-up position).
M12	Cycle stop, after completing block. Push *start* to resume.
M71	Tape rewind. Used unless a ''loop'' tape is used.
M75	Load position, established by the programmer. ''Load'' light goes on, tape and punch are turned off. To resume, push *start,* and program G68.
T01-T20	Random tool selection codes.

* Not used in the program in this book.
SOURCE: Adapted from *Strippit Division, Houdaille Industries, Inc.*

PART DRAWING

The part drawing, Fig. 12-7, is dimensioned for incremental programming. The 0.0, 0.0 location (which is used only in the first punching block) is at the centerline of the punching station.

The "home" position is $X = 1219.20$ mm, $Y = 990.60$ mm [48 × 36 in.] from the centerline of the punch. The machine has an allowable overtravel of 25.4 mm in all four directions.

The signs of the four quadrants (see Fig. 4-1) are the same as in all N/C programs and are viewed, in this machine, by looking from the turret into the clamp. The arrows on the drawing are reversed because they are keyed to the directions of the **workpiece** movements instead of to the relative hole locations.

Tool Codes

Most jobs on a hole puncher use only two to eight different punch shapes and sizes. Thus it is possible to load the 20 stations with an assortment of tooling which will take care of quite a few jobs without any changes. Table 12-3 shows what might be a "standard" arrangement. The pro-

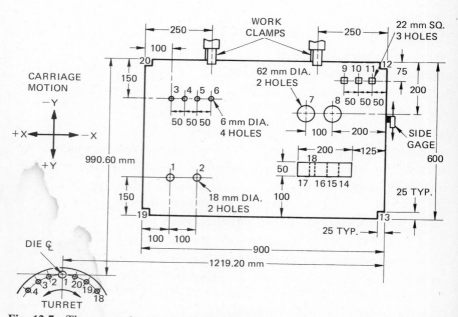

Fig. 12-7 The part to be punched. Dimensioning is not ideal, but fairly good for use with incremental programming. (Adapted from *Strippit Division, Houdaille Industries, Inc.*)

Table 12-3 TOOL FUNCTION COMMAND CODES (HYPOTHETICAL STANDARD TOOLING LOAD)

Tool station	Use	Punch size in.	Punch size mm*	Tool station	Use	Punch size in.	Punch size mm*
1	√	2 × 2	50 × 50	11	√	2.500	62.0
2		0.125	3.0	12		0.437	11.0
3		0.187	4.75	13		1.000	25.0
4		0.218	5.54	14		0.625	16.0
5		0.500† ×1.000	12.0† ×25.0	15	√	0.250	6.0
6		3.000	75.0	16		3.500	88.0
7		0.343	8.72	17		0.500‡ ×1.000	12.0‡ ×25.0
8		0.500	12.0	18		0.937	24.0
9		0.562	14.28	19	√	0.875 square	22.0 square
10	√	0.750	18.0	20		1.125	28.0

* The mm are rounded-off equivalents.
† Station 5 is obround.
‡ Station 17 is rectangular.

grammer checks the tools to be used and, of course, notifies the operator if there are any changes to be made.

The N/C Program

Programming is WORD ADDRESS, VARIABLE BLOCK; trailing zeros are omitted. The arrangement in Fig. 12-8 is only one of several which could be used. The decimal point location shown for X and Y are **only for metric** dimensions. For inch dimensions the decimal point is *three* digits from the right.

N001

G69   Retracts the carriage (which carries the tool clamps) to home position, in the upper right-hand corner when looking into the clamps.

M06   Positions tool station 1 at the punching station. This ensures that the turret indexing will be from a known location.

N002

X0025   Moves the carriage 25 mm to the left. This is not absolutely necessary, but the position indicators on the

PART PROGRAM – N/C TURRET PUNCH						
PART NO. *S-100*		TAPE NO. *3648*			DATE	
MATERIAL *1045 HRS, 4 mm, 600 × 900 mm*					PAGE */* OF */*	
SET WORK CLAMPS AT *250 AND 650 mm*					NO. REQ'D. *4*	
SEQ. NO. NXXX	PREP FUNCT. GXX	X MOVE X±XXXX.XX	Y MOVE Y±XXXX.XX	TOOL NO. TXX	MISC. FUNCT. MXX	REMARKS
n001	*g69*				*m06*	*PUNCH OFF INDEX TO T01*
n002		*x0025*	*y-00094*		*m75*	*LOADING POSITION*
n003	*g68*	*x04192*	*y055*	*t10*		*PUNCH HOLE 1 18 mm DIA.*
n004		*x01*				*HOLE 2*
n005		*x-01*	*y03*	*t15*		*6 mm - HOLE 3 CHANGE TOOL*
n006		*x005*				*HOLE 4*
n007		*x005*				*HOLE 5*
n008		*x005*				*HOLE 6*
n009		*x035*	*y-005*	*t11*		*62 mm - HOLE 7 CHANGE TOOL*
n010		*x01*				*HOLE 8*
n011		*x005*	*y0125*	*t19*		*22 mm SQ. HOLE 9 CHANGE TOOL*
n012		*x005*				*HOLE 10*
n013		*x005*				*HOLE 11*
n014		*x005*	*y0075*	*t01*		*CORNER NOTCH #12*
n015			*y-06*			*CORNER NOTCH #13*
n016		*x-015*	*y0125*			*#14 - 1st PUNCH*
n017		*x-0045*				*#15 - 5mm SHORT*
n018		*x-0045*				*#16*
n019		*x-006*				*#17 - LEAVE A BRIDGE*
n020		*x003*				*#18 CUT OUT BRIDGE*
n021		*x-063*	*y-0125*			*NOTCH #19*
n022			*y06*			*NOTCH #20*
n023	*g67*	*x-03192*	*y-09906*		*(m71)*	*RETURN TO LOAD POSITION*

Fig. 12-8 The N/C program for punching Fig. 12-7 according to one method. All dimensions are in millimetres. (Adapted from *Strippit Division, Houdaille Industries, Inc.*)

clamps are set to read correctly with this amount [1 in.] of *X* offset.

Note: The side gage does not move in the *X* direction, so the 1219.20-mm distance does not change. The side gage is manually moved up while gaging, then lowered below the table. It can be moved in the *Y* direction to make it easier to position wide or narrow sheets.

Y − 00094   Moves the clamps away from the turret 9.40 mm. This, plus the 990.60 normal home dimension, locates the sheet at 1000 mm from zero at this loading position. It is not required, but it makes calculating N003 somewhat easier.

| M75 | Stops the punch and tape while the operator unloads, loads next piece, and closes the clamps. Operator pushes *start* to continue. |

N003   Punch first hole 18 mm in diameter.

X04192   Hole 1 is 900 − 100 = 800 mm from the right side, and 1219.20 − 800 = 419.20 mm move to the left (plus) to center of hole 1.

Y055   Hole 1 is 600 − 150 = 450 mm from the upper edge, and 1000 − 450 = 550 mm move.

T10   Selects the 18-mm punch. Interlocks prevent actual punching until the tools are in position.

N004

X01   is a 100-mm move to hole 2. Carriage moves left so sign is plus. No *Y* move.

N005

X − 01   Moves carriage to the right 100 mm in line with hole 3.

Y + 03   Moves the carriage forward (plus) 300 mm between holes 2 and 3.

T15   Positions the 6-mm punch and die set.

N006, N007, and **N008**

X005   Moves the carriage 50 mm to the left for each of holes 4, 5, and 6.

N009

X035   Moves to hole 7. Hole 6 is 250 mm from the left, and hole 7 is 300 mm from the right. The total 300 + 250 = 550, and 900 − 550 = 350 mm between holes. Table moves left (plus).

Y − 005   As hole 6 is 150 mm and hole 7 is 200 mm from the top, the carriage must move 50 mm up (minus).

T11   Moves the 62.0-mm-dia. punch into position.

N010

X01   Moves the table left (plus) 100 mm to hole 8. No *Y* move needed.

N011

X005   Moves 50 mm to hole 9.

Y0125   Moves the carriage 200 − 75 = 125 mm forward (plus) to hole 9.

T19   Positions the 22-mm-square punch.

N012, N013

X005   Is the 50-mm move between holes 9, 10, and 11.

N014   Notch top right-hand corner.

T01   Is a 50-mm square punch. The *edges* of this punch must be located as shown.

| X005 | As the corner of the part happens to be at the centerline of the 50-mm punch, a 50-mm move is needed. |
| Y075 | As in the X move, a 75-mm move is needed to the centerline of the punch, to notch 12. |

N015

| Y − 06 | Represents a 600-mm move to bring the centerline of the punch to notch 13. |

N016   Start punching the 50 × 200 mm rectangle.

| X − 015 | The centerline of the punch must be at 125 + 25 = 150 mm from the right edge. Carriage moves right (minus). |
| Y0125 | Centerline of the punch must be at 100 + 25 = 125 mm from position 13. Carriage moves down (plus). |

N017, N018

| X − 0045 | Moves the punch 45 mm between positions 14, 15, and 16 instead of the full 50 mm so that no thin strips will be left. This is the same as was done when face milling with several passes. |

N019

| X − 006 | The punch moves all the way to the end, leaving a 10-mm "bridge." Otherwise the punch would be cutting only 10 or 15 mm at its last cut, and this places a bending moment on the punch which might damage it.

     Centerline of punch must be at 200 − 25 + 175 mm from the right of the rectangle. Punch is now at 25 + 45 + 45 = 115 mm. Thus 175 − 115 = 60-mm move to position 17. |

N020

| X003 | Moves centerline of punch to center of the 10-mm-wide bridge. 25 + 5 = 30 mm to position 18. This is 145 mm to the left of the edge of the slot. |

N021   Punch lower left-hand corner notch.

| X − 063 | Punch is now 145 + 125 = 270 mm from the right edge. 900 − 270 mm = 630 mm to move to position 19. |
| Y − 0125 | Punch is still at Y125. Carriage moves out, so sign is minus. |

N022   Punch upper left-hand corner notch.

| Y06 | Total move, full width of part. Carriage moves inward, so it is plus. |

Note: Hole 20 might be programmed first, then 19, so the carriage would be closer to home position. This makes a long "run" from hole 18, but the total movement is shorter.

N023   Move close to load position.

G67   Stops the punch, but not the tape.

X − 03192 Moves 1219.20 − 900 = 319.20 mm to the right, minus.

Y − 09906 Moves 990.60 mm back to original zero point. Move is away from turret (minus). M71 would be used if the tape were on reels and had to be rewound. Omit this if, as in a short 750-mm [30-in.] tape like this, a loop tape is used.

Conclusion

As you can see, incremental programming, especially where there are rows of holes evenly spaced, can be quite simple. While more complex work is frequently done, this program is typical of much of the work done on these versatile N/C hole-punching machines.

Computer programs and CNC (see Chap. 13) are readily available on these machines to speed up the programming, cut down on errors, and aid in troubleshooting and making "on line" corrections.

Chapter 13

USE OF THE COMPUTER IN N/C

There are still many thousands of N/C programs that are relatively simple, and quite short, which are handwritten and typed out on a tape-making typewriter. Thus hand programming is still very much used even in large companies.

However, where contouring is needed, or a large number of holes, or repeated patterns are to be programmed, the use of a computer can literally save hundreds of hours of work. The computer is also about 100 percent accurate—if the proper input is fed to it. Computer people have a motto: GIGO (garbage in, garbage out), which means that the preplanning and the card punching must be correct or the results will be useless.

Before considering the actual computer programming, we will look at some smaller but much discussed uses of the computer for numerical control work.

CNC (Computer Numerical Control)

Since 1970, or a little before, there has been a strong effort to improve the usefulness of the MCU (machine control unit). The relative perfection of the microchips, such as those used in hand calculators, helped this along.

Thus, in early 1975 a new idea started. This idea will, many believe, by 1980, make all present MCU's obsolete. This idea is called CNC (computer numerical control).

CNC has taken some of the abilities of a computer and, in a very compact package, applied them to "stand alone" or "dedicated" service to a single N/C machine. Previously the effort had been devoted to DNC (see later in this chapter) which uses one very large computer to control a lot of machines.

The CNC type of MCU is, as with all previous MCU's, located close (within 1 to 3 m) of the machine. Many CNC's look much the same. However, built into each of these controls is a minicomputer. Minicomputers may have 8K to 32K (K means 1000) "word" capacity, and occasionally more. This capacity is on PCB's (printed circuit boards), each using IC's (integrated circuits) made from "chips" like those used in hand calculators. One quite advanced model is shown in Fig. 13-1.

A CNC control may be very simple and provide only storage for the information on 15 m [50 ft] of tape, or it may have all the features described in this section. The cost may be only $3000 more than a conventional MCU, or it may be $30 000 or more above the standard cost.

The most widely used features are:

1. **Memory Storage**   This is the ability to run the tape once and put it into memory; from then on the program goes to the N/C machine

Fig. 13-1   A complete CNC machine control unit. This includes a CRT and a built-in keyboard for entering data (MDI) and making corrections. (*Cincinnati Milacron.*)

from the computer instead of from the tape. This storage capacity may be from 15 to 210 m [50 to 680 ft] of tape. On some CNC's more than one program can be stored.

This of course, after loading the program, eliminates any further need for the tape. Aside from saving on maintenance of the tape reader, the information comes from the computer much faster, which saves the time lost in relatively slow tape reading.

2. **Edit**   This is the ability to edit, or "debug," the tape right at the machine. It is a fairly expensive addition, but the one which can make the greatest saving.

   Once the tape program is in the CNC minicomputer, changes can be made in the computer's memory without referring to the tape. These changes may be to correct actual location errors in X, Y, or Z, or they may be to optimize the feeds or speeds, or tool path. They are done by MDI (manual data input).

3. **CRT** (Cathode Ray Tube) is like a small, 300-mm [12-in.] TV tube. Words and numbers appear on this tube, often in green. With the CRT, there is a keyboard, usually alphanumeric (that is, with both letters and numbers) for manual data input.

   One type of keyboard is shown in Fig. 13-1. This is built into the control unit. Other controls may use a separate typewriter wired to the CRT for MDI.

   Incidentally, this keyboard in some CNC's can be locked so that only authorized people can make changes to the program.

4. **Corrected tape**   After all debugging has been done, it is possible to produce a corrected tape. This is done with an optional, extra-cost, separate unit which is plugged into the CNC.

5. **Diagnostics**   This refers to troubleshooting when the CNC "goes down." It often takes one or more hours to find the difficulty, and maybe only ten minutes to fix it.

   Two methods of diagnosing trouble are used. The first is with a special diagnostic tape which is supplied with the CNC unit. This tape checks many different paths and, on the CRT, or on a separate oscilloscope, or by means of signal lights, indicates when trouble is found.

   Second, as this tape is limited in its ability to search and signal, the CNC can, by a special telephone unit, be connected directly to the manufacturer's service department. They can run a wider variety of tests, and reportedly up to 90 percent of the time they can spot where the trouble is.

6. In addition, the CNC may take care of inch/metric conversion, tool offsets, ipr/ipm feed rates, feed and speed override, and EIA/ASCII coding. Of course, some of these can also be coded on

Fig. 13-2   A CNC input unit with two disk drives and a minicomputer. (*Cincinnati Milacron.*)

the tape. However, items such as axis-inversions and canned cycles (if they are not available from the tape) can be added by the CNC controller.

A more complex but very compact unit is shown in Fig. 13-2. By means of the keyboard the operator can enter programs or other information onto either of two disks (top two sections) or directly into the minicomputer (bottom unit). The video display unit, or CRT (cathode ray tube), above the keyboard enables the operator to get a visual image of the input, or to read from the disks, for corrections or additions.

When buying a unit, don't be awed by the advertising of a CNC unit. Get very specific details on what is included, its cost, and if additional capabilities can be plugged in later. Some units use additional "software" in the form of tapes or "floppy disks" (magnetic), which can change the CNC's control features, though this is not the most frequently used arrangement.

There is nothing new with CNC insofar as our lessons on programming are concerned. However, it can make many N/C jobs easier, faster, and thus more economical.

DNC (Direct Numerical Control)

Several large companies are using DNC today. They have one or more very large, remote computers, each of which can control up to 40 N/C machines at the same time. This means that 40 different N/C programs are stored in the computer and, by time-sharing, the computer keeps them all going. There is often a CRT and a keyboard at each N/C machine for correction and to give information to the operator. These are usually BTR (behind the tape reader) connections, so the tape reader doesn't even know what's going on. A tape may be made for use in case the computer "breaks down," but they are very seldom used.

As these computers cost $2 000 000 and up, DNC isn't at present used by very many companies, even though, when properly operated, it is the most accurate, least expensive way of operating N/C machines. Companies such as Lockheed and Northrup are using DNC very extensively.

CAM (Computer-Aided Manufacturing)

Our CNC and DNC are really a form of computer-aided manufacturing. However, the most frequent implication is that this system is for collecting management information. This system is connected to each N/C machine and, by cards or typewriter, the operator punches in job numbers. Then the computer keeps track of items such as number of parts made, down time, setup time, and types of machine failures. It can also be used for scheduling and material and inventory control.

At the end of each shift, this information is summarized in the computer and a printout is made for various management people.

As with DNC, only relatively large companies are using this system, though many smaller companies use a modification of it to keep track of productivity.

CAD (Computer-Aided Design)

Since computers were first used they have been employed to do the large amount of mathematics required by some engineering designs. However, CAD today usually implies the use of a computer with a CRT and a keyboard for IO (Input, Output) and often a "light pen," which, when pointed at the CRT, will draw lines or specify changes.

Several companies, especially in the fields of electronics and aerospace, are using this system. One of the most expensive aspects is getting a computer program developed for the specific need. One company reports it took two worker-years to develop its program.

Some programs include computer files of standard shapes of bolts, beams, holes, etc., which can be called for by the designer. Once these are on the screen, the designer can specify their size and location.

Other programs may, for example, allow the designer to put various loads on a beam, and the stress and deflection will appear on the screen. Thus, in aerospace design, many possible loadings can be investigated in a very short time.

Once a data bank has been established (the long preliminary job), design, simulation, analysis, and evaluation of one system, or its variations, can be done fairly rapidly.

In some cases the final design can be transferred to the actual tape, or to a printout of an electrical circuit.

As CAD has been used since the 1960s there are now several programs available. A very large computer is needed for most of these, so the use of CAD is still limited to companies needing sophisticated design capabilities.

USING THE COMPUTER FOR N/C TAPES

Even companies having only two or three N/C machines today find that they need computer assistance for some of the work they do. If they do not own a computer, they can rent time, or hire a specialist who has knowledge of computer programming and access to a computer.

Why Use Computer Programs for N/C Machines?

Timesaving Feature

The drawings used for examples and practice in this book are quite simple. The actual parts machined in industry are frequently more complicated, with many hole locations and milling cuts needed. As it becomes necessary to locate more holes, or a number of holes around a circle, or a pattern of holes frequently repeated, it would be too time-consuming to do all the arithmetic, geometry, and trigonometry by hand.

The more advanced N/C machines can do contour milling on all kinds of curved surfaces (NCC machines). The computations to figure these cuts become quite lengthy. For example, it would be necessary to figure the X and Y values for 62 points to make the required 61 cuts around a 150-mm-dia. semicircle if the tolerance is to be held to $+0.025 - 0.000$.

Accuracy

Doing dozens of computations by hand increases the chance of human errors. The computer will do even very complicated mathematics without making errors. All the programmer does is to get dimensions from the part drawing and tell the computer the arrangement (or pattern) of the holes or cuts to be made.

What Can the Computer Do?

Mathematics

The computer is a very high speed calculating machine. It can figure the X, Y, and Z coordinates of points on a circle or sphere. It can add up a series of dimensions, compute points around any mathematical curve, and perform many other kinds of mathematics. These computations will be accurate to 0.000 000 2 mm [0.000 000 01 in.] if necessary. However, accuracy to 0.0025 mm [0.0001 in.] (or occasionally, [0.00005 in.]) is close enough for N/C work.

Translation

The computer can be programmed, or set up, to understand and translate words such as CIRCLE, TANTO, REV, GOTO, etc., and to use these words to start doing its computations. It can translate the word DRILL into commands to rapidly lower the spindle, feed it slowly, and rapidly retract the spindle. Similarly, translation of MILL, BORE, etc., can be done by the computer.

What Kinds of Computers?

There are several companies now making computers. Some of these have a fairly small "brain," or storage capacity (from 8000 to 40 000 cores, or places to store bits of information). Others have hundreds of thousands of cores. Most of these computers can be used for at least some numerical control work. But complicated jobs require fairly large computers. Some of the available computers and off-line equipment are shown in Figs. 13-3 to 13-6.

If a manufacturing company does not have a computer which can do the needed N/C work, there are computer centers which will sell time on medium- and large-size computers. Thus a manufacturing company can write its own program (or even, in some cases, hire the computer center to write the program) and have the N/C tape made for them at the computer center.

What Do You Feed the Computer?

A Basic Set of Instructions

Computers are completely "stupid"; that is, they cannot "think." Every bit of a computer's amazing skill must be put into it by some person. So it is really the people who have the "amazing skill." Every single step the computer takes in all computation, translation, card punching, etc., must

Fig. 13-3   A large scale computing system used in N/C applications, including APT. A later version of this computer was used for the 336-hole program shown in this chapter. (*UNIVAC Div., Sperry Rand Corp.*)

be specified by the **computer programmer.** This may take hundreds of hours of patient work by highly trained people.

Before a computer can do numerical control, a set of instructions called a **compiler** must be loaded into it which tells it what to do when the commands written by the N/C programmer are fed in.

The N/C Program

After the basic compiler program has been loaded into the computer, we can write our N/C instructions in quite a simple language. These instructions are the N/C program, which is written in a special language, plus dimensions taken from the drawings. The computer "reads" the programs, handles them according to the compiler instructions (previously loaded), and makes all the needed computations for the N/C tape. This process is shown in Fig. 13-7.

The Postprocessor

These processed instructions usually come out of the computer as punched cards or on magnetic tape. They are now in a special code, which is frequently in computer language, and thus very difficult to understand.

As noted in previous chapters, each make and model of an N/C machine is somewhat different from all the others. Thus the point locations

Fig. 13-4   Partial view of one model of a high-speed computer system which is used for N/C (including APT) as well as in product design and research. (*IBM.*)

and other instructions just processed must be interpreted again before they can be punched into the tape.

The machine-tool manufacturers (or the computer companies) therefore make another set of compiler instructions which will change the computer language signals into the kind of coding that a particular N/C machine requires. This set of instructions, shown near the bottom of Fig. 13-7, is called a **postprocessor.** The prefix "post-" means "after." There are dozens of postprocessors, one for each machine, though progress is being made toward creating more "universal" types of postprocessors. A few companies have written computer programs for a specific N/C machine, so that the postprocessor is not needed.

How Is the Tape Made?

The path through the postprocessor to the tape may be one of several. It is possible to have the computer itself punch a tape and even print or

Fig. 13-5   One type of off-line tape converter. Today this could be run immediately from the stored postprocessed computer information. (*IBM*.)

typewrite a copy of what is on the tape, plus added instructions. However, computer time costs from $30 to several hundred dollars an hour, and tape punching, typing, or printing is slow compared with the computer's speed.

Thus the results of postprocessing are usually output in the form of punched cards or coded magnetic tape. The cards, or "mag tape," are then put onto a machine **separate** from the computer (off line, this is called), and this machine makes the actual punched tape or the N/C magnetic tape (see the right-hand path at the bottom of Fig. 13-7 and also Figs. 13-5 and D-6).

What Are These Instructions for Making N/C Tapes?

The instructions are called by many names, depending on which computer company or machine-tool manufacturer developed them.

The names are usually compressed into "words." Some well-known computer programs are given in the following list.

APT (Automatically Programmed Tools)   This is the "top" in N/C com-

Fig. 13-6 The forerunner of CNC, this shows a digital computer connected directly to a Fabrimatic punching machine. This was first used in 1967 and can be used to prepare new tapes while the computer runs one or more punches. (*Strippit Division, Houdaille Industries, Inc.*)

puter programming. It was also the first to be developed. It can be used only on large-capability computers. APT can perform the complicated mathematics of complex curves. Thus some of the N/C programs using this language must be done by programmers who have a good understanding of mathematics. New short cuts and improved methods are constantly being developed, since this is the basic N/C computer program for the entire industry.

The APT language uses over 260 words plus punctuation. It has been estimated that the APT III program represents over one hundred worker-years of development and testing. At present, the further development of APT is in the hands of the Illinois Institute of Technology Research Institute (IITRI), under the APT Long Range Program.

AD-APT (Air Material Command—Developed APT) This is a somewhat limited version of APT which can be run on medium-sized computers. AD-APT uses about 160 words plus punctuation.

AUTOMAP (AUTOmatic MAchining Program) This is a still more limited version of APT which will work with straight lines and circles. It will run on medium-sized computers, and is quite easy to learn. AUTOMAP uses about 50 words plus punctuation.

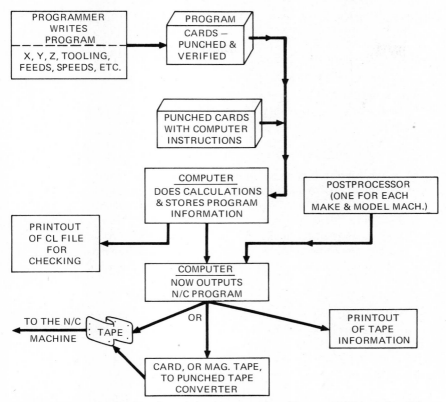

Fig. 13-7 Schematic drawing of the process of making an N/C tape by using a computer.

SPLIT (Sundstrand Processing Language Internally Translated) This program must be run on large computers. No postprocessor is needed if it is used with Sundstrand machines.

SNAP (Simplified Numerical Automatic Programmer) This is a simple, effective program for point-to-point work and can be used on "small" computers. No postprocessor is needed.

AUTOSPOT (AUTOmatic System for POsitioning Tools) Uses about 100 words plus punctuation. Mostly point-to-point work. A good system, though seldom used now.

Is All This Totally Confusing?

Actually, the language used in many of these programs is quite similar. They all use a variety of "pidgin-English" vocabulary which helps the learner considerably. Thus, if one point-to-point and one contouring language were learned, it would not be difficult to learn others if it became necessary to do so.

Who Can Write the N/C Program for the Computer?

The people who write the original sets of instructions (the compilers) for the computer need to be highly trained, and are usually college graduates. These people have done their job so well that their programs are not difficult to put to use.

Point-to-Point N/C Programs (NPC)—Qualifications Needed

These programming languages can be learned by any person who has a sound knowledge of basic algebra, geometry, and simple trigonometry. Some companies and some postsecondary schools teach this type of work. Learning these languages may take from relatively few hours to a few weeks, plus practice in writing and "debugging" the programs.

Of course, a thorough knowledge of good machining practice and of tooling, etc., is needed, just as it is for hand programming.

Contour Programming (NCC)—Qualifications Needed

Some of this kind of programming can be done with the same amount of background as in NPC. Many workpieces which require the use of contour programs can be described by the use of straight lines and circular paths alone. The AUTOMAP and parts of the AD-APT languages are done within these limits. However, parts of the APT and other advanced languages require much more mathematical knowledge, and the wide variety of capabilities of these programs requires considerably more study before a parts programmer can become skilled in using them.

Computer Language

The computer language used for numerical control programming employs many of the same words used in describing any machine shop work. However, in most computers no word over six letters long can be used, so many abbreviations have been made. These abbreviations are common to most N/C computer languages. The computer "understands" the meaning of the following words, and other words, when they are fed in by punched cards or other methods.

Machining Instruction Words

DRILL	SPDRL (spot-drill)
CSK (countersink)	BØREOS (bore, spindle on)
BØRE	REAM
CBØRE (counterbore)	PUNCH

TAP   MILL
FMILL (face mill)

Geometry-Type Instruction Words

ATANGL (at this angle)   TLRGT (keep the tool to the right)
GØLFT (go left)
GØRT (go right)   SETPT (set-point)
GØFWD (go forward)   TANTØ (tangent to)

Machine-Operation Instruction Words

CØØLNT (coolant)   CYCLE (do drilling, etc.)
INDIRV (in the direction   TLØN (tool *on* the line)
  of the vector)   FEDRAT (feed rate)
GØ (move as directed)   KEYBØR (bore, with the tool
GØTØ (move to a point or   prepositioned)
  pattern)

Many programmers also establish a list of abbreviations, as it is time-consuming to write a program using a lot of five- and six-letter words. Thus P or PA for PATERN, PT for PØINT, CI for CIRCLE, LI for LINEAR, etc., can be used, if the computer is told what they are.

With these and other similar words; punctuation; *, ( ), /, $, =, and + and − signs; and some dimensions taken from the part drawing, the computer will do the rest of the computations for even the most complex parts or a large number of hole locations.

Possible Errors

The part programmer, for either manual or computer programming, must be careful that his or her handwritten program is clear and easy for the keypunch operator (the person who punches the cards) to read. For example, the letter O and the number zero look the same when typed or written, but the punched code on the card is quite different. A handwritten 2 may look like a Z, and the letter I may look like the number one. Thus the programmer must always put a top and bottom line on the letter I (or a top hook on the number one), a line through the letter O (Ø) and a line through the center of the letter Z (Z̶). Unfortunately, the O symbol in some computer offices is used for the number zero. One has to ask which system is being used.

CONTOUR PROGRAMS (NCC)

Can Contours Be Hand-Programmed?

If only simple circular cuts and straight lines (horizontal, vertical, or at any angle) are needed to describe the outline of a part, hand programming is not too difficult. However, even these simple parts can require quite a bit of mathematics, which is time-consuming; and it is easy to make mistakes.

For example, to describe the part shown in Fig. 13-8, it would be necessary to compute the points of tangency of the two diagonal lines on the 53.18 R circle, the cutter centerline offset locations, and some tangent cuts around the circle (or use circular interpolation if it is available on the N/C machine). Then you would punch all the coordinates into a tape so that the cutter would mill around the edges.

How Is It Done in APT?

A part as simple as that shown in Fig. 13-8 could even be programmed with AUTOMAP, and the language would be quite similar. The following procedure describes a good method to follow in writing an APT program for Fig. 13-8:

1. **Label all lines and surfaces**
   a. The part in this figure has only a flat surface; so a simple Z location will take the cutter below the bottom edge.

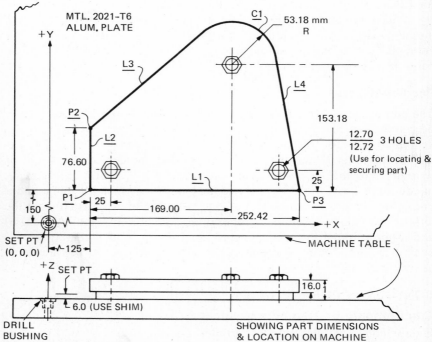

Fig. 13-8   Part and location drawing of a contoured workpiece which is programmed in the APT language.

    b. Each needed point or line or circle is given a simple label, in this case, P1, P2, L1, etc., and C1, the circular arc.

2. **Identify the part** by name and/or number, the postprocessor to be used (N/C machine and control unit), and any other general information, synonyms, etc.

3. **Define all needed points, lines, and surfaces in APT language.** Dimensions are frequently given from a setup point on the part, fixture, or machine table. In APT there are at least 10 possible methods of defining points (and the same number of ways of defining lines and circles); thus a program may be written in several ways, and each method will be correct.

    Notice that several of the definitions in the following discussion do not include any dimensions. They merely describe the geometrical locations in relation to previously defined points, lines, or curves.

    APT words like PERPTO (perpendicular to) and TANTO (tangent to) are easily recognized, as are many of the other words. Of course, not all the APT language is so simple.

4. **Give the computer some information on the cutting tool,** etc. This information may be given before the definitions and may be changed during the program.

5. **Describe the path of the cutter in APT language.** Once the definitions have been established, this can be quite a simple procedure. Sometimes steps 3 and 5 can be combined, using nested defini-

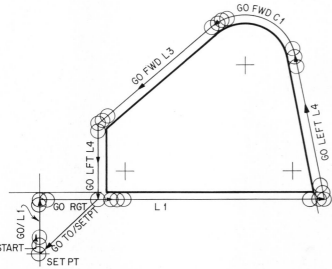

Fig. 13-9 Path of the milling cutter as it cuts the contour of Fig. 13-8.

tions. The method used here is one of the simplest, most easily understood ways of writing a simple APT program. The path followed by the milling cutter is shown in Fig. 13-9.

Along with the cutter-path description there may be some machine commands, such as coolant and spindle control.

THE APT PROGRAM

Step 1   All lines, points, and circles have been given labels in Fig. 13-8.

Step 2   **Identification.**

PARTNØ SAMPLE PART FIG. 13-8
MACHIN/BXCIN, 3 $$ CINCIN.MACH,BENDIX CØNTRØLS
CLPRNT $$ PRINT ØUT ALL CUTTER LØCATIØN PØINTS

Step 3   **Definitions** (Anything to the right of $$ is printed out for information only; it is not part of APT.).

REMARK/ALL DIMENSIONS IN MILLIMETRES

SETPT = PØINT/0,0,0 $$ USE 6 mm SHIM
P1 = PØINT/125,150
P2 = PØINT/125,226.6
P3 = PØINT/377.42,150
L1 = LINE/P1, P3
L2 = LINE/P2, PERPTØ, L1
C1 = CIRCLE/294,303.18,53.18
L3 = LINE/P2, LEFT, TANTØ, C1
L4 = LINE/P3, RIGHT, TANTØ, C1
PL1 = PLANE/0,0,1,0

Step 4   **Tool and other information.**

CUTTER/12 $$ 12 mm END MILL
FEDRAT/300 $$ 300 mm/min FEED
ØUTTØL/0.025 $$ CUT TANGENTS, WITHIN 0.025 OF PERFECT CIRCLE
REMARK/"INTØL" MEANS USE CHØRDS, AND "TØLER"
REMARK/WØULD SPECIFY TANGENTS AND CHØRDS, ØN
REMARK/THE TØØL SIDE ØF THE CØNTØUR

Step 5   **The path, start spindle, etc.**

FRØM/SETPT
INDIRV/0,1,0 $$ $X = 0$, $Y = 1$, $Z = 0$
GØ/L1, PL1
SPINDLE/ØN $$ (sometimes omitted, so that machine operator controls it, for convenience and safety)

CØØLNT/MIST $$ (sometimes also omitted)
TLRGT, GØRGT/L1
GØLFT/L4
GØFWD/C1
GØFWD/L3
GØLFT/L2, PAST, L1
SPINDL/ØFF
CØØLNT/ØFF
GØTØ/SETPT
END $$ END OF A SECTION OF A PROGRAM
FINI $$ END OF A COMPLETE PART PROGRAM

Explanation of APT Program

Many of the statements in the above program are self-explanatory. The notes after the $$ and the REMARK/statements explain several lines. A few additional points should be mentioned.

PØINT, in this program, is defined by giving the X and Y coordinates from SETPT. In other programs SETPT may not be at (0, 0, 0).

LINE is defined, in L1, as going through two previously defined points. However, it should be remembered that a line extends indefinitely on both ends, and may be used beyond the given points. The LINE, L2, extends beyond P2 and below L1, and may be used beyond these locations.

GØFWD is used to continue cutter motion along a tangent path, since no perceptible left or right motion takes place at the tangent point.

LINE, L3, is defined as starting at point P2 and going to circle C1 and TANTO (tangent to) this circle on the left side.

CIRCLE, C1, is defined by three numbers. In APT there are several standard arrangements of numbers. In this circle definition the numbers are *always* arranged as: X to centerline, Y to centerline, radius.

INDIRV means "in the direction of the vector." The following numbers are the $X, Y,$ and Z coordinates from SETPT (as identified in the line above) which establish the *direction* in which the cutter is to travel.

Notice that, with a properly dimensioned drawing, the programmer only had to do a little addition for four definitions. No other dimensional numbers were needed, as the computer did all the mathematics.

As an interesting experiment, you will find that by following the definitions you can draw the entire figure yourself.

A Point-to-Point Program

One of the most dramatic uses of computer programming is to make a tape such as the one for drilling the 336 holes called for in the drawing Fig.

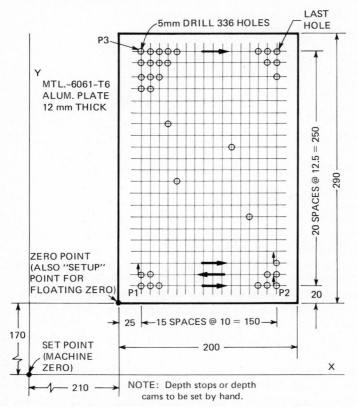

Fig. 13-10   Part and location drawing of plate with 336 holes.

13-10. If hand programming were attempted, the numbers would not be difficult to add, but the time to write all the figures, the time to punch out the tape, and the possibilities of errors in both steps make hand programming a lengthy and difficult job.

The APT Program for 336 Holes

In this example, only three points need to be given labels, as there are several standard routines in APT which can be used for regular patterns of holes. These labels, and all dimensions, are shown in Fig. 13-10.

The following (except for the comments) is copied from an actual computer run for this part. Figure 13-12 shows part of the six pages of CL printout (printout of the centerline locations) which were generated by the computer. The total time, including card handling, was 7.17 min, but the computer took less than three seconds to do the actual calculations. Not all computers are this fast.

Explanation of the APT Terms Used (Fig. 13-11)

Line 2 CLPRNT tells the computer that we want a printout of all the cutter centerline locations after it finishes the calculations.

Line 3 MACHIN is used to identify the N/C machine which will be using the tape. This specifies the postprocessor which is to be used.

Line 4 UNITS/MM specifies that millimetres are used for all dimensions.

Line 5 SETPT specifies where the cutter will be at the beginning of the work. It also, as in floating zero programs, indirectly specifies the zero point the programmer is using.

Lines 6 and 7 Notice that the computer does the simple addition. Points are specified by X, Y, and Z coordinates. The Z is omitted as all work is from one plane.

Lines 9 and 10 PATERN is shown to be located between the two specified points, and the number tells the total number of holes (including the first one) between these points.

1. Identifies the computer and system being used.

```
UNIVAC 1110 TIME/SHARING EXEC---MULTI-PROCESSOR SYSTEM
```

2. Definitions of points and patterns to be used. This is a printout of the 19 punched cards which were fed into the computer.

```
 1.     PARTNO   336 HOLE SAMPLE PROGRAM
 2.              CLPRNT
 3.              MACHIN / TMATIC, 1
 4.              UNITS / MM
 5.     SETPT=   POINT / -210, -170
 6.     P1   =   POINT / 25, 20
 7.     P2   =   POINT / (25 +150), 20
 8.     P3   =   POINT / 25, (20 +250)
 9.     PAT1 =   PATERN / LINEAR, P1, P2, 16
10.     PAT2 =   PATERN / LINEAR, P1, P3, 21
11.     PAT3 =   PATERN / GRID, PAT1, PAT2
12.              PRINT / 0
```

Machining instructions, from punched cards.

```
13.              FROM / SETPT
14.              CYCLE / DRILL, CAM, 1
15.              GOTO / PAT3
16.              CYCLE / OFF
17.              GOTO / SETPT
18.              REWIND / 1
19.              PRINT /0
```

NOTE: Tool specifications, and feeds can also be specified in the program. Our postprocessor does not use this information so it was omitted.

Fig. 13-11 Printout of the 19 cards which were punched out to generate the complete N/C program for drilling 336 holes.

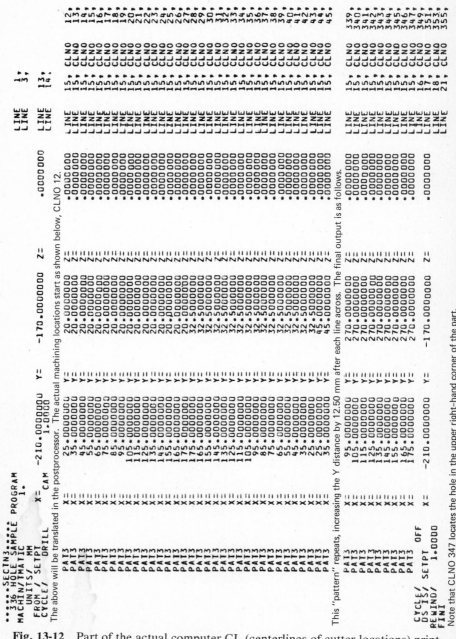

Fig. 13-12 Part of the actual computer CL (centerlines of cutter locations) printout. Notice the zig-zag path.

```
PART NO.—              NAME
336 HOLE SAMPLE PROGRAM
NO TABLE DIAGNOSTICS
001*—21000*—17000        SET POINT    001
002    START                          002
       FEEDRATE    300MM/MIN
       DRILL CHANGE, TOOL NO          003
004*02500*02000       DEPTH22.00      004
005*03500                             005
006*04500                             006
007*05500                             007
008*06500                             008
009*07500                             009
010*08500                             010
011*09500                             011
012*10500                             012
013*11500                             013
014*12500                             014
015*13500                             015
016*14500                             016
017*15500                             017
018*16500                             018
019*17500                             019
020*17500*03250                       020
021*16500                             021
022*15500                             022
023*14500                             023
ETC.
```

Fig. 13-13   Simulation of part of one computer's printout of the tape for a Pratt & Whitney Tape-O-Matic N/C machine using TAB SEQUENTIAL programming.

Line 11   PAT3 = PATERN/. The words PAT1 = etc. specify an arrangement or pattern of holes which can be called from the computer's memory by just the "label." PAT3 uses PAT1 and PAT2 (the horizontal and vertical lines of holes) to form a GRID.

Line 12   PRINT/O merely tells the printer to leave space before printing the next output.

Line 14   CYCLE is the activating word in this program. The words DRILL and CAM, 1 will later go through the postprocessor and generate a g81 and m51 or similar codes.

Line 18   REWIND—this is equivalent to the previous END and FINI. It might, through the postprocessor, generate an m02, or an m30.

Conclusions

These two programs illustrate some of the capabilities of computer programming. As you can see, it isn't very difficult. Though APT is the worldwide standard for N/C programming, many other programs are being used. There are modifications of APT for use on smaller computers.

```
                                      CINTAP POSTPROCESSOR
                                      VERT./HORZ. CINTIMATIC

    N          G                X                   Y              M

PARTNO.    336 HOLE SAMPLE PROGRAM
    1       81              25.0000            20.0000            51
    2                       35.0000
    3                       45.0000
    4                       55.0000
    5                       65.0000
    6                       75.0000
    7                       85.0000
    8                       95.0000
    9                      105.0000
   10                      115.0000
   11                      125.0000
   12                      135.0000
   13                      145.0000
   14                      155.0000
   15                      165.0000
   16                      175.0000
   17                                         32.5000
   18                      165.0000
   19                      155.0000
   20                      145.0000
   21                      135.0000
   22                      125.0000
   23                      115.0000
   24                      105.0000
   25                       95.0000
   26                       85.0000
   27                       75.0000
   28                       65.0000
   29                       55.0000
   30                       45.0000
   31                       35.0000
   32                       25.0000
   33                                         45.0000
   34                       35.0000
   35                       45.0000
   36                       55.0000
   37                       65.0000
   38                       75.0000
   39                       85.0000

         MACHINE TIME THIS FAR =            4.0351 MINUTES

  etc.

         TAPE LENGTH THIS FAR =             3.8500 FEET
```

Fig. 13-14 One computer's printout of the first part of the tape for the 336-hole program for the Cincinnati Milacron N/C machine described in Chap. 6.

For certain machines (especially lathes), the manufacturers have developed special programs which create shortcuts and can be learned in a day or two.

Whatever computer program you use, you will find that they all have quite a lot in common. Thus each one learned makes the next one easier.

A company planning to buy more N/C machines, or more complex

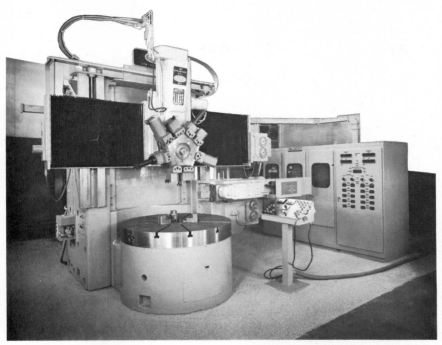

Fig. 13-15 A numerically controlled vertical lathe (VTL). The one shown has a pentagonal (five-sided) turret for deep reach. Used for facing, turning, and boring of straight and contoured shapes. WORD ADDRESS programming; can be manually programmed, but is usually computer-programmed in the APT language. (*Bullard Company*.)

N/C machines, will usually find that computer programming is practically a necessity. Of course, the making of tapes (sometimes several hundred feet long) for complex work cannot be done any other way. A great many simple parts, however, can be programmed more quickly by hand.

Past experience indicates that most companies will find that sooner or later they will be using one or more of the many excellent N/C computer programs for some part of the work of preparing tapes for their N/C machines. Additional information can be obtained from the N/C machine manufacturers, the computer manufacturers, and the technical institutes and colleges that are teaching the subject.

.

Chapter 14

PROS AND CONS OF N/C IN THE SHOP

N/C and Quantity Production

When the price and size of numerical control machines came within reach of smaller plants, the early reaction was, "We don't have the long production runs to pay off the cost."

It was soon evident that this was not the way to get cost savings with N/C equipment. Today, it is realized that the largest savings in N/C are frequently from small lot sizes.

For example, if only two side plates (Fig. 14-1) are needed for a prototype, the part would probably be put on a jig borer, to hold the ±0.06-mm [±0.002-in.] spacing tolerances.

> *Note:* Dual-dimensioning Fig. 14-1 points out one problem in converting inches to millimetres. Many N/C machines are equipped to move in increments of 0.02 mm. This is 0.000 79 in. (close to 0.0008 in.), which is a slightly tighter tolerance but compares closely to the common 0.001 in.
>
> If tolerance is ±0.001 in., the allowance is 0.002 in. In units of 0.02 mm, this could be either 0.04 mm [0.001 58 in.] or 0.06 mm [0.002 36 in.].
>
> In actual practice, the 0.06-mm equivalent, as used in Fig. 14-1 would, in most cases, be a satisfactory conversion. However, if a designer were making the original drawing in metric dimensions, he or she might well use the 0.04-mm allowance, as it is "easy" to get on the metric dials of the machines.

If an N/C drill is available, a manuscript could be written and a tape made for this part in about twenty minutes. Locating, aligning, and securing the piece on the machine would take the same time on the jig borer as on the N/C machine. From then on, the tape-controlled machine would locate all the holes within ±0.02 or 0.03 mm [±0.001 in.] in little more

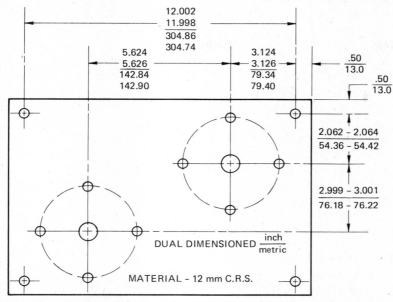

Fig. 14-1 Bearing side plate, 12 mm thick. Dual dimensioned, with metric equivalents, some rounded off.

than the actual time it takes to spot-drill and drill through. Not all tape-controlled machines will work so close, but most will. This could take as little as 10 percent of the time it would take by conventional means, and the two pieces would be identical within about ±0.01 mm [±0.0005 in.].

If a few pieces are needed for a prototype, the numerical control machine will thus frequently be faster and less expensive. If a production run is large, conventional methods may be less expensive than numerical control.

How many is "large"? A figure frequently mentioned is 300 pieces. However, some runs of more than 1000 pieces have been economical on tape-controlled equipment. Also, some runs of 100 pieces might be more economical by conventional methods if tooling is already available.

If accuracy of hole location would ordinarily require drill jigs or jig boring, and if the quantity is from 1 to 100 pieces, numerical control will probably save considerable time and money.

Lower Tooling Costs

By taking advantage of the built-in accuracy of numerical control, it is often possible to simplify greatly the tooling needed for a job. This means a saving in cost of tooling, and a saving in lead time, which is often equally important. This economy may be due to the ellimination of drill jigs, since the N/C machine will locate points within ±0.13 to ±0.02 mm [±0.005 to ±0.001 in.] consistently.

Further savings result if a part has drilling and milling cuts. Conventional machining will require jigs for drilling and a fixture for milling. However, the numerical control machine can often do both types of operation on one setup. This, of course, increases accuracy.

Saving of tooling costs is especially noticeable on the N/C machines known as machining centers. These often have a rotary table and, with proper fixturing, can machine four or five surfaces of a part at one setup. Ordinarily, this would require several fixtures.

Of course, it is necessary to design and make some kind of holding fixture for machining work on numerical control equipment. Since the fixtures are for holding only, however, they are frequently quite simple and inexpensive.

Easier Design Changes

Especially during the development of new equipment, and often at other times, the engineering department specifies changes in design. With conventional machining this can require scrapping or reworking of the tooling previously used, which is expensive and delays the work. If the change is on a part being made with numerical control equipment, it is usually only necessary to change the tape, a job requiring sometimes only a few minutes, and at most a few hours.

In fact, if holes or milling are to be added to a part, a new section of tape can be made and spliced into the present tape.

If the changes are in the **sizes** of holes or milling cuts, no change in the tape may be needed. The operator may be able to use different cutting tools and regulate feeds and speeds at the console until a new tape is ready without stopping production.

Simplified Inspection

Once the first piece has passed inspection, not much checking is necessary on the output from a numerical control machine. The **locations** of all holes and cuts are assured by the tape and the machine.

Of course, depths of milling cuts, diameters of bored or reamed holes, and fits of tapped holes still must be checked. These items, however, can be checked quite quickly with simple gages. The lengthy process of checking hole locations need only be done at infrequent intervals.

An interesting result attached to the importance of the first piece inspection is the development of equipment to speed up this job. Careful checking of the center-to-center locations of several holes is a rather slow

Fig. 14-2   A coordinate measuring machine (CMM) which can check X and Y absolute or incremental dimensions to the nearest 0.0025 mm [0.0001 in.]. Newer models have inch/metric capabilities. Z axis can also be checked, and automatic printout is available. (*Sheffield Corp.*)

job, and time on an N/C machine may be worth from \$15 to over \$100 an hour. Therefore inspection machines called coordinate measuring machines (CMM), such as that shown in Fig. 14-2, have been developed by several companies. These machines show the dimensions in lighted numbers (and printed on a tape if desired) to the nearest 0.02 or 0.002 mm [0.001 or 0.0001 in.] and work as fast as the inspector can move the stylus.

Reduced Scrap

With the tape controlling most of the important factors in the machining operation, the human factor has less effect on final accuracy. The errors due to operator fatigue, interruptions, etc., are much less likely to happen. Of course, on the simpler numerical control machines, the operator changes tools and manually sets the feeds and speeds. Thus the quality of the operator's work still can affect the final quality of the product.

   If the numerical control machine is completely automatic, the operator's principal responsibility is to load the part properly and inspect each part for finish, hole size, tapped hole size, etc., to make sure that he or she changes or adjusts the cutting tools soon enough.

The overall result of fewer, simpler fixtures, more machining without moving the part, and less chance of human error has been to greatly reduce the percentage of scrap or reworked parts when they are made on N/C machines instead of with conventional equipment.

Reduced Space Requirements

The most immediate space-saving results from fewer jigs and fixtures needing to be stored. In fact, one large company changed to N/C for part of its production of replacement and repair parts and reduced jig and fixture storage from 7000 m² [75 000 sq ft] to just under 930 m² [10 000 sq ft]. The tapes must be stored, but a 100- to 300-mm [4- to 12-in.] square by 32-mm [1¼-in.] high box takes a lot less space than most jigs or fixtures.

The locating, transporting, returning to storage, maintenance, and repair of jigs and fixtures are all time-consuming jobs which are decreased when their number is reduced.

Improved Production Planning

When work is done on a numerical control machine, the time to load and cut the part is very nearly constant, so that scheduling can be more precise. In addition, the N/C machine often performs, at one setting, work which formerly was scheduled through several conventional machines, each with a different machine load.

Because N/C is economical for smaller lots, it is often not necessary to keep as large a parts inventory on hand. Thus less storage space is needed and more flexible production schedules are possible.

Parts Particularly Suited to N/C Machines

In summary, the machining of a part on numerical control equipment should seriously be considered if one or more of the following conditions apply:

1. Parts would require expensive jigs or fixtures if machined on conventional equipment.
2. Parts for which maximum use can be made of N/C with quick-change tools or automatic tool changers to perform a number of different cuts on a workpiece.
3. Parts require close tolerance and/or good repeatability.
4. Parts may need several engineering changes, such as experimental or prototype equipment.
5. Human error would be especially costly.
6. Parts are needed in a hurry.
7. Small lots, short production runs are planned.

N/C Is Not a Cure-all

Today, up to 50 percent of the new machines purchased by both large and small companies have some form of numerical control. The use of numerically controlled equipment is even reaching into spot-welding setups and assembly of parts. However, no one believes that N/C will ever totally replace conventional equipment for the following reasons:

N/C does not—
    Totally eliminate errors
    Eliminate the need for careful fixture design (though there will be fewer and often simpler fixtures)
    Reduce the cost of high-quantity, long-run, well-tooled jobs
    Eliminate the need for good inspection procedures
    Enable a tool to cut much faster than it can on conventional equipment

Before buying numerical control equipment, management must take into consideration that:

N/C does—
    Mean greater original cost of machine
    Require better and more skilled mechanical, electrical, and electronic maintenance personnel
    Often require the use of preset cutting tools, thus extra planning and careful tool grinding
    Affect drafting and inspection procedures, especially in the change-over period
    Require retraining of machine operators and training of a new group of programmers

Thus, like any advanced machining technique, numerical control must be used only where it will do the work better or faster or more accurately than other methods. However, many shops have found that the savings and advantages have been greater than they expected. Therefore anyone working in the metal-removal or metal-shaping field should become familiar with the wide range of prices, types, and capabilities of numerical control equipment.

USEFUL FORMULAS AND TABLES

I. **Cutting speed** is specified in surface feet per minute (sfpm or, more often, just fpm) or in metres per minute (m/min).

In the following equations:

dia. = diameter, in **inches** or **millimetres,** of the rotating cutter, or the diameter of the surface being cut on a rotating workpiece

Cutting speed equations are:

In inches

$$(A)\ \text{fpm} = \frac{(\text{rpm})(\text{dia.})}{4}$$

$$(B)\ \text{rpm} = \frac{(4)(\text{fpm})}{\text{dia.}}$$

Note: The constant 4 is an approximation of $12 \div \pi = 3.82$.

In metric

$$\text{m/min} = \frac{(\text{rpm})(\text{dia.})}{300}$$

$$\text{rpm} = \frac{(300)(\text{m/min})}{\text{dia.}}$$

Note: The constant 300 is an approximation of $1000 \div \pi = 318$.

These constants are sufficiently accurate for practically all machine shop work.

II. **Feed.** Feed for N/C machines must often be expressed in inches per minute (ipm) or in millimetres per minute (mm/min).

Feed from the published tables is, however, given in inches per revolution (ipr) or in millimetres per revolution (mm/rev). Milling feeds in the tables are in inches per tooth (ipt) or in millimetres per tooth (mm/t).

For drills, reamers, countersinks, and lathes:

In inches
(A) Feed (ipm) = (rpm)(ipr)

In metric
Feed (mm/min) = (rpm)(mm/rev)

For all milling cutters:

(B) Feed (ipm) = (rpm)(ipt)
× (number of teeth)

Feed (mm/min) = (mm/t)
× (number of teeth)(rpm)

For tapping and threading:

(C) Feed (ipm) = (rpm) ÷
(threads/in.)

Feed (mm/min) = (rpm)(pitch)

III.   **Metal Removal.** Rate of metal removal is usually expressed in cubic inches per minute (cim) or in cubic centimetres per minute (cm³/min), though mm³ or m³ could be used in the metric system.

For drills, the area of the cross section (cutting area) is $\pi d^2/4$, where d is the diameter in inches or millimetres.

$$(A) \; \text{cim} = (\text{area of drill})(\text{rpm})(\text{feed, ipr})$$
$$= (\text{area of drill})(\text{feed, ipm})$$

$$(B) \; \text{cm}^3/\text{min} = \left(\frac{\text{area of drill}}{100}\right)\left(\frac{\text{feed, mm/rev}}{10}\right)(\text{rpm})$$

$$= (\text{area of drill})(\text{feed, mm/min}) \div 1000$$

For milling cutters—linear cutting (plunge cutting with end mills, use same formula as for drills). All dimensions in inches or millimetres.

$$(C) \; \text{cim} = (\text{width of cut})(\text{depth of cut})(\text{rpm})(\text{feed, ipt})$$
$$\times (\text{number of teeth})$$
$$= (\text{width of cut})(\text{depth of cut})(\text{feed, ipm})$$

$$(D) \; \text{cm}^3/\text{min} = \left(\frac{\text{width of cut}}{10}\right)\left(\frac{\text{depth of cut}}{10}\right)(\text{rpm})$$

$$\left(\frac{\text{feed, mm/t}}{10}\right)(\text{number of teeth})$$

$$= (\text{width of cut})(\text{depth of cut})(\text{feed, mm/min}) \div 1000$$

For lathes—all dimensions in inches or millimetres. The millimetres are changed to centimetres by dividing by 10, and metres per minute × 100 = cm/min. These cancel out in the metric formula.

$$(E) \text{ cim} = (\text{depth of cut (feed, ipr)}(\pi \times \text{dia.})(\text{rpm})$$
$$= (\text{depth of cut})(\text{feed, ipr})(\text{fpm} \times 12)$$
$$(F) \text{ cm}^3/\text{min} = (\text{depth of cut})(\text{feed, mm/rev})(\text{m/min})$$

IV. **Horsepower needed.** A quick though approximate method of figuring the horsepower, or kilowatts, needed for a cut is to multiply the volume of metal removed in a minute (cim or cm^3/min) by a constant P (which has been determined by experiments) and divide the result by the machine efficiency. Values of P vary according to the source

Table A-1 UNIT POWER REQUIREMENTS AT 100% EFFICIENCY*

Material	Horsepower per cubic inch per minute† = P			Kilowatts per cubic centimetre per minute† = P metric		
	Drilling	*Milling*	*Turning*	*Drilling*	*Milling*	*Turning*
Mild steel to 25 Rc	1.0	1.0	0.9	0.0456	0.0456	0.041
Med. steel 25–30 Rc	1.6	1.8	1.3	0.073	0.082	0.059
Hard steel 35–50 Rc	1.9	2.1	1.5	0.0866	0.0956	0.0683
Soft cast iron	0.8	0.7	0.5	0.0365	0.032	0.0228
Hard cast iron	0.9	1.1	1.0	0.041	0.050	0.0456
Aluminum	0.35	0.4	0.3	0.016	0.0183	0.0137
Brass	0.5	0.6	0.4	0.0228	0.0274	0.0183
Bronze	0.6	0.8	0.7	0.0274	0.0365	0.032
Stainless steel						
-400 series	1.3	1.3	1.1	0.059	0.059	0.05
-300 series‡	1.6	1.8	1.7	0.073	0.082	0.0775
Titanium	1.0	1.0	1.1	0.0456	0.0456	0.05
Nickel alloys	1.6	1.6	1.5	0.073	0.073	0.0683

Note: Heavier feeds require less unit hp. Double these feeds will need about 20% less hp.

* Values given here are approximately the average of values from several sources. Individual values from reliable sources may vary from these values.
† Conversion factors: 1 in.^{3} = 16.39 cm^3; 1 hp = 0.746 kW.
‡ Free machining alloys 30–40% less.
Adapted from Roberts and Lapidge, *Manufacturing Processes.* New York: McGraw-Hill Book Co., 1977.

of information used. It will also vary somewhat with changes in cutting-tool geometry, feed rate, depth of cut, and cutting speed. However, the figures in Table A-1 will result in reasonable estimates. The formula is:

(A) hp = (cim)(P) ÷ efficiency
(B) kW = (cm³/min) (P metric) ÷ efficiency

Example: Drilling C1018 mild steel with a 12.7-mm [¹/₂-in.] drill at 23 m/min [75 fpm], with 0.10 mm/rev [0.004 ipr] feed, at 70 percent efficiency, and using Table A-1:

$$\text{rpm} = \frac{(300)(23)}{12.7} = 543 \text{ rpm} \qquad \text{(Eq. I}B\text{)}$$

$$\text{cim} = \left[\frac{\pi \, (0.5)^2}{4} \right] (543)(0.004) = 0.43 \text{ cim} \qquad \text{(Eq. III}A\text{)}$$

$$\text{hp} = (0.43)(1.0) \div 0.70 = 0.61 \text{ hp required} \qquad \text{(Eq. IV}A\text{)}$$

In metric:

$$\text{cm}^3/\text{min} = \left[\frac{\pi \, (12.7)^2}{4 \times 100} \right] (543) \left(\frac{0.10}{10} \right) = 6.88 \text{ cm}^3/\text{min}$$
$$\text{(Eq. III}B\text{)}$$

kW = (6.88)(0.0456) ÷ 0.70 = 0.448 kW   (Eq. IV*B*)
Check: 1 hp = 0.746 kW, so 0.746 × 0.61 hp = 0.45 kW

V.   **Circles** ($\pi = 3.1416$)

(A) Circumference = $\pi \times 2R = \pi \times$ dia.

(B) Area = $\pi R^2 = \dfrac{\pi D^2}{4}$

For chords: If the angle A and the radius R are known (Fig. A-1),

(C) Chord length = $C = 2 \left(R \sin \dfrac{A}{2} \right)$

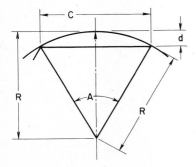

Fig. A-1 A chord (length C) of a circle.

$$(D) \quad \text{Offset} = d = R \left(1 - \cos \frac{A}{2}\right)$$

If the tolerance d and the radius R are known (Fig. A-1),

$$(E) \quad C = 2\sqrt{d(2R - d)}$$

$$(F) \quad \text{Angle } A = \arctan \frac{\sqrt{d(2R - d)}}{R - d}$$

Note: "Arctan" means "the angle whose tangent equals" the numerical value of the expression shown.

For tangents: If the angle A and the radius R are known (Fig. A-2),

$$(G) \quad \text{Tangent length} = T = 2R \left(\tan \frac{A}{2}\right)$$

$$(H) \quad \text{Offset} = d = R \; \frac{1 - \cos \dfrac{A}{2}}{\cos \dfrac{A}{2}} = R \left(\frac{1}{\cos \dfrac{A}{2}} - 1\right)$$

If the tolerance d and the radius R are known (Fig. A-2),

$$(I) \quad T = 2\sqrt{d(2R + d)}$$

$$(J) \quad \text{Angle } A = 2 \arctan \frac{\sqrt{2d(R + d)}}{R}, \text{ or } 2 \arccos \frac{R}{R + d}$$

For secants: If the angle A and the radius R are known, and $d = d'$ (Fig. A-3),

$$(K) \quad \text{Secant length} = s = 4R \sqrt{\frac{1 - \cos \dfrac{A}{2}}{1 + \cos \dfrac{A}{2}}}$$

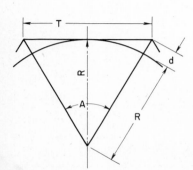

Fig. A-2 A tangent (length T) of a circle.

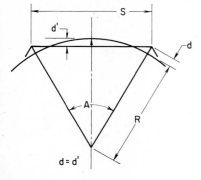

Fig. A-3 A secant (length S) of a circle.

$$(L) \quad d = R \; \frac{1 - \cos \dfrac{A}{2}}{1 + \cos \dfrac{A}{2}}$$

If the tolerance ($d = d'$) and the radius R are known (Fig. A-3),

$$(M) \quad S = 4\sqrt{Rd}$$

$$(N) \quad \text{Angle } A = 2 \arctan \frac{4\sqrt{Rd}}{R - d}, \text{ or } 2 \arccos \frac{R - d}{R + d}$$

Angular Tolerance

Angles are often specified as $\pm \frac{1}{2}°$ tolerance. This seems like a very small amount. Table A-2 shows what this amounts to, in actual tolerance in inches. For comparison, the table also shows the length of the chord if the allowable tolerance is only $\pm 0°5'$, or $\frac{1}{12}°$ (Figs. A-4 and A-5).

The distance t for other angular tolerances will be very close to exactly proportional. For example, for a tolerance of $0°10'$, multiply the t for $5'$ by 2, or divide the t for $\frac{1}{2}°$ ($30'$) by 3.

Table A-2

Radius		Chord length = tolerance			
		$\pm\frac{1}{2}°$		$\pm 0°5'$	
Inches	Millimetres	Inches	Millimetres	Inches	Millimetres
1	25.4	± 0.009	± 0.23	± 0.0015	± 0.038
3	76.2	0.025	0.64	0.0044	0.112
6	152	0.052	1.32	0.0088	0.224
10	254	0.087	2.21	0.0146	0.371
15	381	0.131	3.33	0.0219	0.556

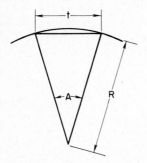

Fig. A-4 Sketch for computing approximate angular tolerance.

The formula is (Fig. A-4)

(P) Tolerance = chord = t = $\pm 2R \sin \dfrac{A}{2}$

Note: When the angle B is given on a drawing (Fig. A-5) and the tolerance is ± angle A, the actual difference in chord length, t, varies somewhat according to the size of angle B. However, Table A-2 serves to indicate the approximate deviation which can be expected.

The exact formula is (Fig. A-5)

$$(Q) \text{Tolerance} = 2R \sin (B \pm A) - 2R (\sin B)$$
$$= 2R [\sin (B \pm A) - \sin B]$$

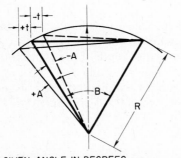

B = GIVEN ANGLE IN DEGREES
A = ANGULAR TOLERANCE ALLOWED, IN
    DEGREES.

Fig. A-5 Sketch for computing actual angular tolerance.

VI.   **Right triangles.** A right triangle is a triangle which has one right angle (90° angle). A common notation system for triangles is to use capital letters for the vertices or angles and small letters to represent the sides opposite the angles, as shown in Fig. A-6. Angle C is 90°, and the hypotenuse is the side opposite the right angle, which is side c. This is the triangle most frequently used in machine-shop calculations.

(A) Angles $A + B + C = 180°$

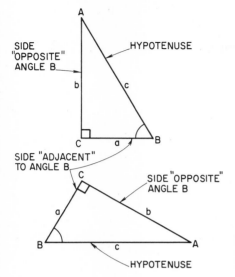

Fig. A-6 Right triangles, showing the names of the sides as used in trigonometry.

Since angle $C = 90°$ by definition, then

(1) Angles $A + B = 90°$

Therefore

(2) Angle $A = 90° -$ angle B

and

(3) Angle $B = 90° -$ angle A

(*B*) The famous Pythagorean theorem proved that the lengths of sides of a right triangle were related as follows: *The square of the hypotenuse equals the sum of the squares of the other two sides:*

(1) $c^2 = a^2 + b^2$ (Fig. A-6)

Thus

(2) $c = \sqrt{a^2 + b^2}$

(3) $a = \sqrt{c^2 - b^2}$

(4) $b = \sqrt{c^2 - a^2}$

(*C*) It has been proved that in a right triangle the ratios between the lengths of the three sides depend on the sizes of the angles A and B (angle C in our illustration is always $90°$). These ratios are only for right triangles. They are called trigonometric, or "trig," functions, and they are published in many textbooks and handbooks.

The ratios have the following names and definitions, referring to Fig. A-6, and using angle B as the reference angle, so that side b

is "opposite" and side a is "adjacent." "The angle" in trig formulas is always formed by the hypotenuse and one side.

1.* $\text{Sine} = \sin = \dfrac{\text{opposite}}{\text{hypotenuse}}$

2.* $\text{Cosine} = \cos = \dfrac{\text{adjacent}}{\text{hypotenuse}}$

3.* $\text{Tangent} = \tan = \dfrac{\text{opposite}}{\text{adjacent}}$

4. $\text{Cotangent} = \cot = \dfrac{\text{adjacent}}{\text{opposite}}$

5. $\text{Secant} = \sec = \dfrac{\text{hypotenuse}}{\text{adjacent}}$

6. $\text{Cosecant} = \csc = \dfrac{\text{hypotenuse}}{\text{opposite}}$

Notice that if, in a **right** triangle, one side and one angle are known, or two sides are known, all the other sides or angles can be calculated by using the formulas above and the trig tables. If angle A were used as our reference angle, then side b would be the adjacent side, and side a the opposite in the above formulas.

VII.   **Drill-point allowance.** The formula for the length of the point, P, for a drill, countersink, or other pointed tool is (Fig. A-7)

$$(A)\ P = \frac{d}{2}\left[\tan\left(90° - \frac{A}{2}\right)\right] \qquad \text{where } d = \text{diameter}$$

By finding, in the tables, the value of the tangent, dividing this by 2, and calling this number K, the formula above can be simplified to

$$(B)\ P = (\text{dia.})(\text{constant}) = dK$$

The value for this constant for several point angles is shown in

A = POINT ANGLE
d = DIAMETER
P = LENGTH OF THE POINT

Fig. A-7 Sketch for computing the point length of a drill.

* These three functions (or formulas) are by far the most frequently used.

Table A-3 VALUES OF K FOR DRILL
POINT ANGLES

Point angle	K	Point angle	K	Point angle	K
60°	0.87	118°*	0.30	135°	0.21
82°	0.58	120°	0.29	145°	0.16
90°	0.50	125°	0.26	150°	0.13

* 118° is the standard drill point angle.

Table A-3. It is usually sufficiently accurate to interpolate for the constant for angles not listed in the table.

VIII. For convenience in working problems when using this book, Tables A-4 and A-5 give the tap drill sizes of the most used threads. The tap drills are for the 75 percent threads, which are still the most frequently used.

In metric taps, the ''M'' number is the bolt diameter in millimetres, and the second number is the pitch in mm. There is no metric equivalent of our ''threads per inch.''

Table A-4
INCH SIZES

Tap size	Tap drill
8–32	#29
10–32	#21
$1/4$–20	#7
$5/16$–18	F
$3/8$–16	$5/16$
$1/2$–13	$27/64$
$5/8$–11	$17/32$
$3/4$–10	$21/32$
1–8	$15/16$

Table A-5
METRIC SIZES

Tap size	Tap drill
M4 × 0.7	3.3
M5 × 0.8	4.2
M6 × 1.00	5.0
M8 × 1.25	6.8
M10 × 1.50	8.5
M12 × 1.75	10.3
M16 × 2.0	14.0
M20 × 2.5	17.5
M24 × 3.0	21.0

MAGIC-THREE CODING FOR SPEEDS AND FEEDS

The Electronic Industry Standards formerly recommended that this coding system be used. It is a three-digit coded number, though provisions were also made for four- or five-digit coding if more accuracy was necessary.

Most N/C machines made since 1974 do not use this coding as they have direct inches per minute or millimetres per minute coding or require simply the use of a table for speeds.

However, since several hundred N/C machines which use the "magic three" are still being used, it is shown here.

The "three" in this coding means two things:

1. The code is made up of only three digits, plus a letter code if WORD ADDRESS is being used.
2. "The first digit of the coded number is a decimal multiplier and has a value three (3) greater than the number of digits to the left of the decimal point," to quote EIA.

Item 2 uses one of the three available digits, which means that the rpm or ipm must be rounded off to two digits. In most cases this is sufficiently accurate.

For example, 75.7 rpm has two digits to the left of the decimal point. The first digit in the code is therefore $3 + 2 = 5$. The 75.7 would be rounded off to 76, and the magic-three code for 75.7 rpm is 576, and might be programmed as S576.

If there are no digits to the left of the decimal point, subtract the

Table B-1

Feed or speed	Magic-three coding*	Four-digit† coding
1728.	717	7173
150.	615	6150
15.2	515	5152
7.82	478	4782
0.153	315	3153
0.0126	213	2126
0.00875	188	1875
0.000462	046	0462

* The magic-three coding is used on some N/C machines, though many other coding systems are also used, as noted in the preceding chapters.
† Shown in AIA and NAS Standards only.
> *Note:* The second digit can never be zero unless all digits are zero.

number of zeros *after* the decimal point from 3 to get the first number of the code.

For example, a feed of 0.0757 ipm has one zero to the right of the decimal. Thus the first code digit is $3 - 1 = 2$, and the 0.0757 feed is coded 276 and might be programmed as f276.

A list of examples, as shown in the former EIA and AIA Standards, is given in Table B-1.

ZERO OFFSET

ZERO OFFSET is used on most fixed-zero N/C machines today, in order to shorten the setup time. It is sometimes referred to as ZERO SHIFT, or FULL ZERO SHIFT, and is incorrectly called FLOATING ZERO in some machine descriptions.

Both ZERO OFFSET and FLOATING ZERO types of control allow the machine setup person to position the workpiece at any reasonable location on the machine table and then establish or adjust the zero location.

The difference is, as noted in the Glossary (Appendix G), that a true FLOATING ZERO machine has no fixed reference point (or zero point) on the machine table; thus the control retains no information on the location of any previously established zeros. However, the ZERO OFFSET is used on fixed-zero N/C machines and the control retains information on the location of the "permanent" zero. In a FLOATING ZERO control, the zero is "established"; in a ZERO OFFSET control, the zero is moved, or shifted.

The ZERO OFFSET feature may allow an adjustment of only ± 0.13 0.25 mm, to assist the operator in the final steps of "zeroing in." As more frequently used today, the zero point may be shifted or offset to any point on the machine table. This FULL ZERO SHIFT feature is sometimes an extra cost option.

To see how the FULL ZERO OFFSET works, first consider a setup procedure with no ZERO SHIFT available. The part programmer instructs the operator to set up the workpiece, as shown in Fig. C-1. The operator (or setup person) dials in, under manual control setting, the X150.0 and Y125.0 on the decade switches and presses the *cycle start* button. The table (or column) moves so that the center of the spindle is located above

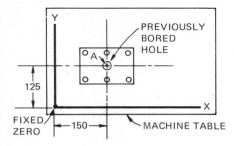

Fig. C-1 · Setup of part without using ZERO OFFSET.

the (150, 125) point on the machine table. The operator must now clamp the work on the table and push and tap the part and tighten and loosen clamps until hole A "trams" in as centered on the spindle within the allowed tolerance, and the workpiece is aligned with the table.

If the FULL ZERO OFFSET is used, the operator locates the workpiece in any convenient place on the machine table, as shown in Fig. C-2. He or she aligns the edges, dials X150.0 and Y125.0, and pushes the *cycle start* button; and the center of the spindle moves, as before, to point A, which is now some unknown distance away from the previously bored hole on which the spindle must be centered.

The operator now uses his ZERO SHIFT dials, which might look like those shown in Fig. C-3. Notice that there is one dial for each axis, X and Y, and a round button underneath, which is for locking in the final setting. The dials may be rotated either clockwise (to move the table toward larger plus values) or counterclockwise (to move the table toward smaller plus values).

The settings of (150, 125) on the manual decade switches of the X and Y readings are **not changed** when the ZERO OFFSET movements are made, since the ZERO SHIFT dials merely operate the table drive motors, and their signal does not enter the memory section of the MCU.

By turning these ZERO SHIFT dials, the operator now moves the table zero into hole A, but at location A'. The dials are usually marked off in

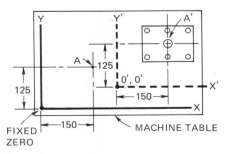

Fig. C-2   Setup procedure when using FULL ZERO OFFSET.

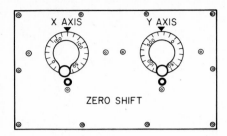

Fig. C-3 One type of ZERO SHIFT dial. One complete turn of a dial moves the table 5.08 mm [0.20 in.].

0.02-mm divisions, so that this adjustment can be made quickly and accurately.

Once the workpiece or fixture is zeroed in, the ZERO SHIFT dials are locked, and this part, and any more of the same part, can now be machined according to the tape commands, based on the $X = 150$ and $Y = 125$ dimensions from the fixed zero to the center of hole A.

Notice that none of this affects the work of the programmer. Its only purpose is to save time in setting up the N/C machine.

TAPE READERS
AND DUPLICATORS

The majority of numerical control machines use either an electromechanical or a photoelectric tape reader. Both types are accurate, efficient, and reliable. The two principal differences are in the cost and the speed of reading the tape.

Electromechanical Tape Readers

For point-to-point and straight-cut N/C machines, the electromechanical tape reader is most frequently used. This equipment is relatively inexpensive, simple to service, and fast enough to handle all NPC work very easily.

These tape readers "read" the tape at speeds of from 20 to 120 characters (or rows) per second (1200 to 7200 rows per minute). The supply and take-up reels and tensioning bars are sometimes mounted on a plate with the reader, as in Fig. D-1, or sometimes the reels are mounted separately, but of course close to the reader.

On one well-known electromechanical tape reader, the tape is driven for reading and rewinding, and kept aligned, by a rotating sprocket which engages each sprocket hole in the tape. Rewinding is usually at the same speed, which is equivalent to up to 18.3 m/min [60 fpm]. This is quite adequate when you realize that 9.1 m [30 ft] of tape contains 3600 characters or rows, which, at an average of 15 characters per block, means a 240-line program. This is a fairly long program in point-to-point work, though quite short for contouring.

The actual operation of one make of electromechanical tape reader that is very simple is shown in Fig. D-2.

253

Fig. D-1 An electromechanical reversible tape reader. Operates at up to 60 characters per second, using 150 to 250 mm [6 to 10 in.] diameter reels. Faster models are available. (*Tally Corp.*)

There are eight fixed C-shaped contacts having 0.635-mm [0.025-in.] wide openings between the two points. Eight very sensitive movable contacts are set to touch one side of the fixed contact when the star wheel senses "no hole," and the other when a "hole" condition is sensed. Wiring from all three terminals is connected to the machine control unit.

The movable contacts are allowed to drop when the star wheel enters a punched hole. This closes a circuit to the machine control unit, telling it that a hole is present in a certain track. The REWIND, REWIND STOP, and similar codes are interpreted in the machine control unit, and the MCU then operates switches which reverse or stop the tape-reader drive motor.

Other makes of electromechanical tape readers will, of course, have different details of construction. However, the basic principles of the operation will be similar to the foregoing description.

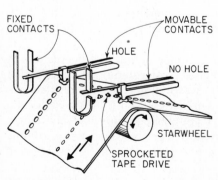

NOTE: Two contacts shown. Eight are used for reading N/C tapes.

Fig. D-2 Detail of star wheels and contacts used in the reading head of the tape reader shown in Fig. D-1. (*Tally Corp.*)

Photoelectric Tape Readers

A contouring-type N/C machine may have to make hundreds of very short cuts when it is approximating a curved surface by means of close-tolerance chords, tangents, or secants. The time needed to make each cut may be very short. For example, a cut 0.25 mm long, made at 250 mm/min feed rate, takes only 0.060 sec (60 msec). The electromechanical tape readers cannot always read the next block of information in so short a time, and a very expensive N/C machine might have to wait for the tape reader.

This problem is solved by adding a buffer storage, or memory, and using a high-speed photoelectric tape reader. A buffer storage is a section of the MCU where a complete block of information may be stored in memory. The procedure is that while the N/C machine is making a cut, a tool change, etc., the tape reader operates, reads the next block, and sends the information into the buffer-storage section. When the machine calls for its next instruction, the instruction goes (with the speed of electricity) from the buffer section into the MCU, and the machine almost instantly starts the next cut or tool change, or whatever the command is. At the same time the tape reader operates and starts the cycle again.

This alone would not solve the problem; so photoelectric tape readers are used on practically all NCC machines. These readers are more expensive, but they operate at speeds from 100 characters (rows) per second up to 1000 rows per second, which is equal to 6000 to 60 000 rows per minute, or 15 to 150 m [50 to 500 ft] of tape per minute if they operate continuously!

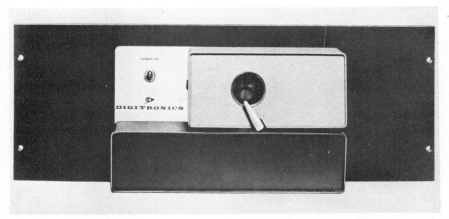

Fig. D-3 A high-speed solid-state photoelectric N/C tape reader. Reading speeds up to 300 characters per second. Made in unidirectional and bidirectional models. (*Digitronics Corp.*)

The basic operation of a photoelectric tape reader is quite simple. For example, by moving the lever shown in Fig. D-3 to the load position, the tape can be slid into this make of reader. Figure D-4 shows the same reader with the top cover removed. A small exciter lamp (sometimes 10 watts output) shines downward, through any holes that have been punched into the tape, and on to one or more of nine photodiode, or silicon solar-cell, heads. A photodiode is like one-half of a transistor which responds to the energy in light. It changes light energy into electric energy in much the same way as the familiar selenium photocell, but a photodiode is much smaller.

The small current generated by the photodiode is fed through amplifiers to make it strong enough to be used in the circuits of the machine control unit.

The light which shines through the **sprocket** hole goes into a photodiode, which is used to gate the signal from the eight data-output circuits. That is, the timing of the length of the data signals is controlled by the sprocket lamp. The need for this is obvious when you notice that the punched code holes in a tape are 1.83 mm [0.72 in.] in diameter and spaced only 2.54 mm [0.100 in.] apart. The sprocket holes are only 1.17 mm [0.046 in.] in diameter, which makes read and nonread signals more nearly equal in duration. This control also helps to assure that the signals from all eight tracks in the punched tape are read at the same time, so that the MCU will read the signal correctly.

The tape readers do not have any drive sprockets; so the movement of the tape is controlled by a rotating capstan, sometimes with a pinch roller underneath. In some tape readers, when the pinch roller rises up, it presses the tape onto the capstan, and the tape is driven. When a stop

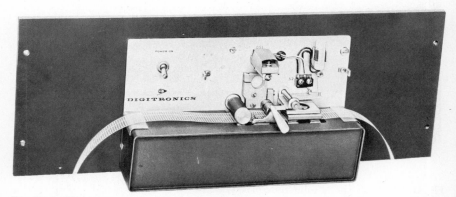

Fig. D-4 The tape reader of Fig. D-3 with the top cover removed, showing the exciter lamps, capstan drive, and brake. (*Digitronics Corp.*)

signal, such as EOB, is read, the pinch roller drops, and the brake presses on the tape and stops it. Some other tape readers use a direct-drive capstan, with a printed-circuit motor. This type of motor does not have an iron rotor, so that it can accelerate or stop in a few milliseconds, and does not use a brake.

Tape Handling

Tape to be fed through the tape readers is wound on reels of various sizes. The reels range in size from 150 mm [6 in.] or less up to 400 mm [16 in.] in diameter, the most frequently used being from 190 to 267 mm [7^1/$_2$ to 10^1/$_2$ in.] in diameter. The length of the tape on the reel depends on the thickness of the tape. The EIA standard tape is 0.094 to 0.109 mm [0.0037 to 0.0043 in.] thick. However, tapes of from 0.051 to 0.114 mm [0.002 to 0.0045 in.] thick can be purchased.

The 267-mm [10^1/$_2$-in.] dia. standard reel is rated to hold 360 m [1200 ft] of 0.114 mm [0.0045-in.] thick tape. It will, of course, hold over twice as much 0.051-m [0.002 in.] thick tape.

The tape-handling equipment must make provision for rewinding the

Fig. D-5 A high-speed tape handler. Bidirectional, with speeds up to 700 characters per second. (*Digitronics Corp.*)

tape and also for compensating for the effects of inertia when the tape starts and stops. The inertia of start-up and the slack-off of stop are frequently absorbed by one or more "dancer" rolls. Tape-handling equipment also keeps the tape aligned, and brings it into the tape reader at the proper angle.

Fig. **D-6** A highly sophisticated punched-tape handler. This machine will duplicate or verify a tape. Working with a computer it can transfer information either way. (*Univac Div., Sperry Rand Corp.*)

The unit in Fig. D-1 shows simple tape-handling equipment mounted on the same panel with the tape reader. Figure D-5 shows a high-speed tape handler which would probably be mounted just below a tape reader.

Tape Duplication

When tapes are punched manually on a tape-punching typewriter, a duplicate can be made on the same equipment. It is also possible to verify or check tapes on these special typewriters. However, the tapes used for large contouring work are usually made from computer programs and may have 610 m [2000 ft] of tape on a reel.

Special equipment has been developed for duplicating and verifying these long tapes. One such unit is shown in Fig. 3-5. This equipment is quite fast (60 to 120 characters per second), and can be set so that it will omit any DELETE codes or blank spaces in the original tape. Parity checking of the newly punched tape may also be accomplished. Another, even more versatile unit is shown in Fig. D-6.

When a company is using many long tapes, the use of the tape-duplicating equipment allows the computers and the tape typewriters to be used only for the more productive work.

SPEEDS AND FEEDS FOR DRILLING, REAMING, TAPPING, MILLING, AND TURNING

The tables of speeds and feeds included in this appendix were selected as a result of searching through many sources: handbooks from the ASTME and the Metal Cutting Tool Institute and publications of the steel, aluminum, etc., industries. The figures given in various sources sometimes vary considerably. The values listed here are those which most closely represent approximate agreement.

Classifications of materials in general terms such as "aluminum" and "brass" are far from precise, since there are many alloys of both. Detailed information on a particular alloy can be found in handbooks published by the respective industries.

Several factors must be considered when deciding whether to use the high or low end of the ranges of speeds and feeds:

1. Rigidity of the machine, tooling, cutter, and workpiece
2. Finish required—rough or finish cut
3. Smooth or interrupted cut
4. Cutting fluids—kind, amount, efficiency
5. Tolerances required, in location and dimensions of cut
6. Consideration of economical tool life or maximum production as being of greater importance
7. Depth of hole or depth of cut, and feed rate
8. Any heat-treatment which has been given the material
9. Number of pieces per production run

The tables can serve the student well as an indication of the relative speeds and feeds of different kinds of materials and as a guide to the variations in feeds and speeds which are necessary because of the differences in the cutting processes of drilling, milling, etc.

For heavy roughing cuts, use the slower cutting speeds. For light finish cuts, use the higher speeds. End mills cannot usually remove as

Table E-1 SUGGESTED CUTTING SPEEDS USING HIGH-SPEED (HSS) CUTTING TOOLS

Material	Drilling		Reaming		Turning	
	fpm	m/min*	fpm	m/min*	fpm	m/min*
Aluminum	250–600	75–180	100–300	30–90	400–1000	120–300
Brass, free cutting	150–300	45–90	130–200	40–60	225–350	70–110
Bronze, soft	100–250	30–75	75–180	23–55	150–225	45–70
Cast iron						
Soft	75–150	23–45	60–100	18–30	100–150	30–45
Medium	70–110	20–35	35–65	10–20	75–120	23–35
Hard	60–100	18–30	20–55	6–17	50–90	15–27
Copper	60–100	18–30	40–60	12–18	100–200	30–60
Magnesium	300–650	90–200	150–350	45–110	600–1200	180–360
Stainless steel						
Free-machining	65–100	20–30	35–85	10–25	100–150	30–45
Other	15–50	5–15	15–30	5–9	40–85	12–25
Steel						
Free machining	100–145	30–45	60–100	18–30	125–200	40–60
Under 0.3 carbon	70–120	20–35	50–90	15–27	75–175	23–50
0.3 to 0.6 carbon	55–90	17–27	45–70	14–20	65–120	20–35
Over 0.6 carbon	40–60	12–18	40–50	12–15	60–80	18–25
Titanium	30–60	9–18	10–20	3–6	25–55	8–17
Zinc die casting	200–400	60–120	125–300	40–90	300–1000	90–300

* Conversions rounded off.
Adapted from Roberts and Lapidge, *Manufacturing Processes*. New York: McGraw-Hill Book Co., 1977.

Note: Carbide tools may be run 2 to 4 times as fast as shown above. See Table E-3 for milling speeds and feeds.

much metal per minute as face mills; so lower speeds should be used. Table E-1 gives suggested cutting speeds for several materials. See Table E-3 for speeds and feeds for milling.

Feeds for Drilling—HSS Drills

Feed rate allowable for drills varies more as a function of drill diameter than of the material being cut. However, the finish is dependent on feed; so feeds smaller than suggested may be used for better finish. Horsepower used also increases with feed rate (though not proportionately); so feeds in the lower end of the suggested range must ordinarily be used for the difficult-to-machine metals, though too fine feeds can result in work-hardening.

Solid-carbide and carbide-tipped drills should be fed at approximately the same feeds as HSS drills, or slightly slower. One reason is that the thrust (downward pressure) increases almost in direct proportion to the feed rate, and too high a thrust could break the drill. Table E-2 gives reasonable drilling feed rates. To change ipr to ipm, see Appendix A.

Feeds for Reaming

A good starting point is to ream at twice the feed rate used for drilling. The range of recommendations is from 1½ to 3 times the drilling feed rate. Finish required, amount of stock being removed, accuracy needed, and tool life will determine whether to use the low or high end of the allowable range for a given-size reamer.

Speeds and Feeds for Counterbores and Countersinks

These two tools have rather light cutting edges, and are usually run at about one-half the speed of the same-size drill.

Some experimenting may be needed to find a speed at which the countersink will not chatter.

Table E-2 FEEDS FOR DRILLING—HSS AND CARBIDE DRILLS

Drill diameter		Drill feed	
Inches	Millimetres	Inches per revolution	Millimetres per revolution
Under ¹/₈	−3.2	0.001−0.003	0.03−0.08
¹/₈−¹/₄	3.2−6.35	0.002−0.005	0.05−0.13
¹/₄−¹/₂	6.35−12.7	0.004−0.007	0.10−0.23
¹/₂−1	12.7−25.4	0.007−0.017	0.18−0.43
Over 1	25.5 up	0.015−0.030	0.38−0.76

Feed rates for the counterbore are about half those for drilling. Countersinks may run at half to three-quarters the feed rate of a similar-size drill.

Feeds for Tapping—HSS Taps

The feed in inches per revolution (ipr) of taps is fixed by the threads per inch (tpi). Every revolution of the tap will advance it one full thread. For metric taps there is no equivalent of "tpi" as the pitch is specified directly in millimetres. So feed rates are:

$$\text{Tap feed (ipr)} = \frac{1}{\text{threads/inch}} = \text{pitch of thread}$$

Tap feed (mm/r) = pitch of thread in mm

Tapping Speeds

Tapping speeds should be set at one-quarter to one-third the speeds shown in Table E-1 for turning. However, on N/C machines, the manufacturer's recommended rpm for tapping is from 200 to 400 rpm even if the resulting m/min is very low.

Feeds for Milling—HSS Cutters

A milling cutter is only as strong as each individual tooth. Thus the important consideration is how big a "bite" or chip each tooth takes as the milling cutter advances. So, in milling the basic feed unit is millimetres per tooth (mm/t). This must be changed to millimetres per minute (mm/min) for use by an N/C machine. The formula is II*B* in Appendix A.

Published figures for milling feeds vary widely since there are so many variables influencing a milling cut. In Table E-3 heavy feeds are used when cutting the free-machining materials, such as aluminum and magnesium. Low feed rates are used with some stainless steels and high-carbon steels. Carbide cutters will run at about the same feeds per tooth as HSS cutters.

Feeds for Turning—Carbide Cutters

Cutting with single-point tools on a lathe is quite different from drilling, reaming, etc. Lathe-turning toolholders can be made large in cross section, with carbide inserts, so that very heavy cuts may be taken (up to the usable horsepower of the lathe) without breaking the cutting tool.

As the feed increases, the power required and the heat generated also increase. To compensate for these factors the rpm can be decreased. The SME *Tool and Manufacturing Engineers Handbook* (McGraw-Hill Book Company, New York) has a table showing how feed, depth of cut, and

Table E-3 END MILLS—SPEEDS AND FEEDS

| Material | Feed per tooth | | | | Cutting speed | | | |
| | 6.35-mm [1/4-in.] dia. end mill | | 25-mm [1-in.] dia. end mill | | Roughing | | Finishing | |
	Inches per tooth	Millimetres per tooth	Inches per tooth	Millimetres per tooth	Feet per minute	Metres per minute*	Feet per minute	Metres per minute*
Aluminum	0.003	0.08	0.009	0.23	600	180	800	245
Bronze, med.	0.003	0.08	0.007	0.18	250	75	300	90
Bronze, hard	0.002	0.05	0.005	0.13	125	40	150	45
Cast iron, soft	0.003	0.08	0.008	0.20	60	18	80	25
Cast iron, hard	0.002	0.05	0.005	0.13	50	15	70	20
Plastic, glass-filled	0.003	0.08	0.012	0.30	150	45	160	50
Steel								
Low carbon	0.001	0.03	0.004	0.10	75	23	90	30
4140	0.005	0.013	0.003	0.08	50	15	70	20
4340	0.0003	0.008	0.002	0.05	50	15	70	20
Stainless steel								
Type 304	0.001	0.03	0.004	0.10	55	17	75	23
Type 17-4PH	0.0005	0.013	0.003	0.08	35	10	50	15
Inconel	0.0002	0.005	0.003	0.08	30	9	40	12
Monel	0.0003	0.008	0.004	0.10	60	18	80	25
Ti-6Al-4V	0.001	0.03	0.004	0.10	25	8	40	12
Zinc die casting	0.002	0.05	0.010	0.25	800	240	1000	300

Note: FACE MILLS use up to twice the above feed rates. For carbide cutters, double the above cutting speeds.
* Conversion rounded off

Table E-4 FEEDS FOR ROUGH TURNING—HSS OR CARBIDE CUTTERS

Material	Feed rate	
	Inches per revolution	*Millimetres per revolution*
Aluminum	0.007–0.050	0.18–1.27
Cast iron	0.011–0.025	0.28–0.64
Copper alloys	0.005–0.022	0.13–0.56
Nickel alloys	0.005–0.018	0.13–0.46
Stainless steel	0.005–0.022	0.13–0.56
Steel	0.010–0.090	0.25–2.30
Titanium	0.007–0.018	0.18–0.46

cutting speed can be varied. Feeds from 0.05 to 2.3 mm/rev [0.002 to 0.090 ipr] are shown in this SME tabulation for a wide variety of metals.

Table E-4 gives a range of feeds as recommended by reliable sources. Many shops will occasionally use roughing feeds greater than those shown here, sometimes using circular-shaped carbide inserts.

Feeds for finish turning are usually from 0.04 to 0.26 mm/rev [0.0015 to 0.010 ipr] depending on the finish desired. Studies have been made proving that, with a larger nose radius, fine finishes can be achieved with fairly high feed rates.

EXCERPTS FROM NATIONAL CODES

The following excerpts are from some of the national codes, printed with permission from Electronic Industries Association, 2001 Eye Street, N.W., Washington, D.C. 20006; and Aerospace Industries Association of America, Inc., 1725 De Sales Street, N.W., Washington, D.C. 20036. Another source of information is the former United States of America Standards Institute, which is now the American National Standards Institute, 1430 Broadway, New York, N.Y. 10018. The latter organization publishes standards on N/C tape, coding, etc., which are the same as those from EIA. They also publish standards on many other subjects, from electronics to grinding wheels. Complete copies of the codes and standards may be purchased from the above associations.

The EIA standards shown in the next few pages are the latest revised edition available at the time of this writing.

Axis and Motion Nomenclature
for Numerically Controlled Machines

(From EIA Standard RS-267 and Standards Proposal No. 916, formulated under the cognizance of EIA Engineering Committee TR-31 on Numerical Control Systems and Equipment.)

1. Scope

This standard for axis and motion nomenclature for numerically controlled machines is intended to simplify programming, to simplify the training of programmers, and to facilitate the interchangeability of control tapes.

1.1 This standard applies to all numerically controlled machines.

1.2 Definitions of terms used in this standard are in accordance with EIA Automation Bulletin 3B. Glossary of Terms for Numerically Controlled Machines,'' February, 1965, or latest revision thereof.

1.3 Revisions to these standards will be proposed as technical progress warrants.

2. The Standard Coordinate System

2.1 The Standard coordinate system gives the coordinates of a moving tool (or draftman's pencil) with respect to a stationary workpiece.

2.2 It shall be used for numerical control programming, machine fixturing, and machine loading.

2.3 It shall be designated by unprimed letters such as x, y, and z. (See paragraph 10.)

3. The Z Axis of Motion

3.1 The z axis of motion is parallel to the principal spindle of the machine. If there are several spindles, one shall be selected as the principal one. If there is no spindle the z axis is perpendicular to the work holding surface. If the principal spindle can be swiveled or gimbaled the z axis is parallel to the spindle axis when the spindle is in its 0 position. The preferred 0 position is with the spindle perpendicular to the work holding surface.

3.1.1 On such equipment as milling, boring, drilling, and tapping machines, the spindle is the tool rotating means.

3.1.2 On such equipment as lathes, grinders, and other machines which generate a surface of revolution, the spindle is the work rotating means.

3.2 Positive z (w and r, see paragraph 8) is in the direction from the workholding means toward the tool-holding means. Positive z motion increases the distance between the work and the tool. (See paragraph 10.)

4. The X Axis of Motion

4.1 The x axis of motion is horizontal and parallel to the workholding surface.

4.2 If z is horizontal, positive x is to the right looking from the spindle toward the workpiece.

4.3 If the z axis is vertical, when looking from the spindle toward its supporting column(s) the positive x axis is to the right on single column machines or forward on dual column or gantry machines.

4.4 On machines generating a surface of revolution, such as lathes, x, u, and p motions shall be radial, and normally the positive direction of motion shall be away from the center of revolution. Where the linear motion can cross the centerline of rotation, positive motion shall be in the direction of maximum displacement from the center of rotation.

5. The Y Axis of Motion

5.1 The y axis of motion is perpendicular to both x and z.

5.2 Positive y is in the direction to make a right-handed set of coordinates; i.e., + x rotated into + y advances a right-handed screw in the + z direction.

6. Rotary Motions, A, B, and C

6.1 a, b, and c are angles defining rotary motions around the axes parallel to x, y, and z respectively.

6.2 Positive a, b, and c are in the directions to advance a right-handed screw in the + x, + y, and + z directions respectively.

7. The Origin of the Standard Coordinate System

7.1 The location of the origin (x = 0, y = 0, z = 0) of the standard coordinate system and/or the angular origins (a = 0, b = 0, c = 0) may be fixed or adjustable.

8. Additional Axes

8.1 If in addition to the primary linear slide motions, x, y, or z, there exists secondary slide motions parallel to these, they shall be designated u, v, and w respectively. If tertiary motions exist, they shall be designated p, q, and r respectively. If linear motions exist, which are not or may not be parallel to x, y, or z, they may be designated u, v, w, p, q, or r, as is most convenient. If there are more than three (3) sets of parallel motions, unused letter addresses may be used in alphabetical order.

8.2 If in addition to the primary rotary motions a, b, or c, there exist secondary rotary motions, either parallel to a, b, or c or compound or gimballed to a, b, or c, they shall be designated d or e.

8.3 The primary linear motions are those nearest the principal spindle; the secondary motions are those next nearest; the tertiary are the farthest. For example, the carriage on a turret lathe is z, while the saddle, being farther from the spindle, is w.

8.4 It is recommended that axis selection be based upon the most complex version of a machine. Simpler versions of such a machine

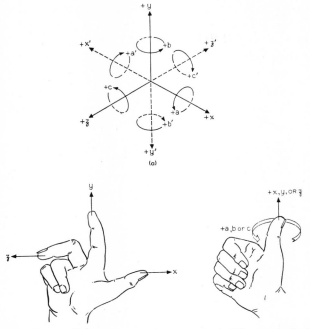

Fig. F-1 Right-hand coordinate system. (*Electronic Industries Association.*)

derived by a deletion of assigned axes need not alter the designation of positive directions on the remaining axes.

9. Direction of Spindle Rotation

9.1 Clockwise spindle rotation is in the direction to advance a right-handed screw into the workpiece.

10. Reversed Directions for Moving Workpieces

10.1 If a machine element moves the workpiece instead of the tool, it must respond to the tape in the opposite direction from that defined above for moving the tool. In illustrating various machine tools, an arrow with a primed letter, such as + x′, is the direction of motion of a moving workpiece for a command calling for positive motion; while an arrow with an unprimed letter, such as + x, is the direction of motion for the same positive command of the tool with respect to the workpiece. (The programmer, fixturing set-up man, and machine loader should think exclusively in terms of the unprimed directions. Only the machine builder need consider the primed directions.)

11. Schematic Drawings of Numerically Controlled Machines (Fig. F-2)

11.1 Schematic drawings of typical machines interpreting this standard are shown on the following pages.

11.1.1 The schematic drawings indicate by letters and arrows the motions normally numerically controlled. On machines similar to those shown, but with more or less numerically controlled motions, Section 8 shall apply.

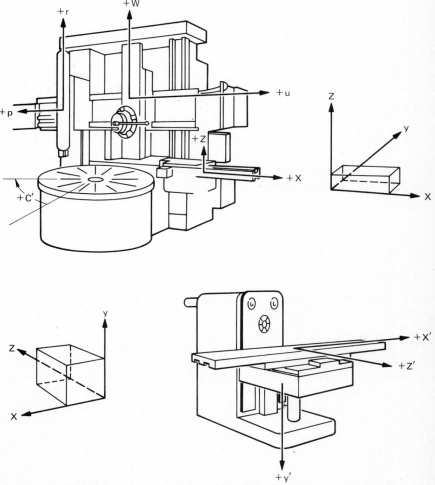

Fig. F-2 Axis motion designation, according to EIA RS-267-A as applied to two frequently used N/C machines.

Interchangeable Perforated Tape Variable Block Format for Positioning, Contouring, and Contouring/Positioning Numerically Controlled Machines

(From EIA standard RS-274-C, replacing EIA Standards RS-274-B, RS-273-A, and RS-326-A and Standards Proposals 1136 and 1147, formulated under the cognizance of EIA Committee TR-31 on Numerical Control Systems and Equipment.

4. Addresses

4.1 The address character shall be in accordance with Table F-1.

    4.1.1 Where one "Feed Function" word is used, the address shall be "f". Where additional "Feed Function" words are used, their addresses shall be in accordance with Table F-1.

    4.1.2 Where one "Rapid Traverse Dimension" word is used, the address shall be "r". Where additional "Rapid Traverse Dimension" words are used, their addresses shall be in accordance with Table F-1.

Note: Additional commands may be required on specific machines. Unassigned code numbers should be used for these and specified on the Format Classification Sheet. On certain machines the functions described may not be completely applicable; deviations and interpretations should be clarified in the Format Classification Sheet. (See Appendix A.)

Table F-1

Character* RS-244	RS-358	Address for
a	A	Angular dimension around x axis
b	B	Angular dimension around y axis
c	C	Angular dimension around z axis
d	D	Angular dimension around special axis or third feed function or tool function for selection of tool compensation† ‡
e	E	Angular dimension around special axis or second feed function†
f	F	Feed Function
g	G	Preparatory Function
h	H	Permanently unassigned
i	I	Interpolation parameter or thread lead parallel to x
j	J	Interpolation parameter or thread lead parallel to y
k	K	Interpolation parameter or thread lead parallel to z

Table F-1 (*Continued*)

l		Do not use
	L	Permanently unassigned
m	M	Miscellaneous Function
n	N	Sequence Number
o	⋮	Reference Rewind Stop
	O	Do not use
p	P	Third rapid traverse dimension or tertiary motion dimension parallel to x†
q	Q	Second rapid traverse dimension or tertiary motion dimension parallel to y†
r	R	First rapid traverse dimension or tertiary motion dimension parallel to z†
s	S	Spindle Speed Function
t	T	Tool Function
u	U	Secondary motion dimension parallel to x†
v	V	Secondary motion dimension parallel to y†
w	W	Secondary motion dimension parallel to z†
x	X	Primary x motion dimension
y	Y	Primary y motion dimension
z	Z	Primary z motion dimension
End of Record	%	Rewind Stop
Carriage Return	Line Feed	End of Block
Tab	Tab	Tab
+	+	Plus
−	−	Minus
/	/	Block Deleter (slash)
	)	Control In
	(	Control Out

* RS-244 is EIA code, and RS-358 is ASCII code.
† Where d, e, p, q, r, u, v, and w are not used as indicated, they may be used elsewhere.
‡ On drafting machines, the character "d" may be used as a second miscellaneous function word with the following assignments:

d1—Stylus Down	d4—Dash Line Generator "on"	d7—Unassigned
d2—Stylus Up	d5—Dash Line Generator "off"	d8—Unassigned
d3—Index Turret	d6—Unassigned	d9—Transfer Hold

Table F-2   PREPARATORY FUNCTIONS

Code	Function
g00	Point to Point, Positioning
g01	Linear Interpolation
g02	Circular Interpolation Arc CW
g03	Circular Interpolation Arc CCW
g04	Dwell
g05	Unassigned
g06	Parabolic Interpolation
g07	Unassigned
g08	Acceleration
g09	Deceleration
g10–g12	Unassigned
g13–g16	Axis Selection
g17	XY Plane Selection
g18	ZX Plane Selection
g19	YZ Plane Selection
g20–g24	Unassigned
g25–g29	Permanently Unassigned
g30–g32	Unassigned
g33	Threadcutting, Constant Lead
g34	Threadcutting, Increasing Lead
g35	Threadcutting, Decreasing Lead
g36–g39	Permanently Unassigned
g40	Cutter Compensation/Offset, Cancel
g41	Cutter Compensation-Left
g42	Cutter Compensation-Right
g43	Cutter Offset, Inside Corner
g44	Cutter Offset, Outside Corner
g45–g49	Unassigned
g50–g59	Reserved for Adaptive Control
g60–g69	Unassigned

Table F-2 PREPARATORY FUNCTIONS (*Continued*)

g70	Inch Programming
g71	Metric Programming
g72–g79	Unassigned
g80	Fixed Cycle Cancel
g81	Fixed Cycle No. 1
g82	Fixed Cycle No. 2
g83	Fixed Cycle No. 3
g84	Fixed Cycle No. 4
g85	Fixed Cycle No. 5
g86	Fixed Cycle No. 6
g87	Fixed Cycle No. 7
g88	Fixed Cycle No. 8
g89	Fixed Cycle No. 9
g90	Absolute Dimension Input
g91	Incremental Dimension Input
g92	Preload Registers
g93	Inverse Time Feedrate (V/D)
g94	Inches (millimeters) per Minute Feedrate
g95	Inches (millimeters) per Spindle Revolution
g96	Constant Surface Speed, Feet (meters) per Minute
g97	Revolutions per Minute
g98–g99	Unassigned

1. Permanently unassigned codes are for individual use and are not intended to be assigned in future revisions of the standards.
2. Assignments in previous revisions:

*	g05	Hold
*	g10	Linear Interpolation (Long Dimension)
*	g11	Linear Interpolation (Short Dimension)
*	g20	Circular Interpolation Arc CW (Long Dimension)
*	g21	Circular Interpolation Arc CW (Short Dimension)
*	g30	Circular Interpolation Arc CCW (Long Dimension)
*	g31	Circular Interpolation Arc CCW (Short Dimension)
*	g45–g49	Cutter Compensation
†	g60–g79	Reserved for Positioning

* These are now "Unassigned."
† g60–g69 and g72–g79 are now "Unassigned."

Table F-3 MISCELLANEOUS
FUNCTIONS

Code	Function
m00	Program Stop
m01	Optional (Planned) Stop
m02	End of Program
m03	Spindle CW
m04	Spindle CCW
m05	Spindle OFF
m06	Tool Change
m07	Coolant No. 2 ON
m08	Coolant No. 1 ON
m09	Coolant OFF
m10	Clamp
m12	Unassigned
m13	Spindle CW & Coolant ON
m14	Spindle CCW & Coolant ON
m15	Motion +
m16	Motion −
m17–m18	Unassigned
m19	Oriented Spindle Stop
m20–m29	Permanently Unassigned
m30	End of Tape
m31	Interlock Bypass
m32–m35	Unassigned
m36–m39	Permanently Unassigned
m40–m45	Gear Changes if Used; Otherwise Unassigned
m46–m47	Unassigned
m48	Cancel m49
m49	Bypass Override
m50–m89	Unassigned
m90–m99	Reserved for User

Notes: Permanently unassigned codes are for individual use and are not intended to be assigned in future revisions of the standard.
Assignments in previous revisions: m32–m35 Constant Cutting Speed, and m36–m39 Unassigned.

Table F-4 EXPLANATIONS OF FUNCTIONS (partial list only)

g00 Point to Point Positioning	Point to point positioning at rapid traverse rate along random path.
g01 g01 Linear Interpolation	Contouring control which uses the A mode of contouring control which uses the information contained in a block to produce a straight line in which the vectorial velocity is held constant.
g02 Arc Clockwise (Arc CW) See Circular Interpolation	An arc generated by the coordinated motion of two axes in which curvature of the path of the tool with respect to the workpiece is clockwise, when viewing the plane of motion in the negative direction of the perpendicular axis.
g03 Arc Counterclockwise (Arc CCW) See Circular Interpolation	An arc generated by the coordinated motion of two axes in which curvature of the path of the tool with respect to the workpiece is counterclockwise, when viewing the plane of motion in the negative direction of the perpendicular axis.
g02–g03 Circular Interpolation	A mode of contouring control which uses the information contained in a single block to produce an arc of a circle. The velocities of the axes used to generate this arc are varied by the control.
g04 Dwell	A timed delay of programmed or established duration, not cyclic or sequential; i.e., not an interlock or hold.
g41 Cutter Compensation-Left	Cutter on left side of work surface looking from cutter in the direction of relative cutter motion with displacement normal to the cutter path to adjust for the difference between actual and programmed cutter radii or diameters.
g42 Cutter Compensation-Right	Cutter on right side of work surface looking from cutter in the direction of relative cutter motion with displacement normal to the cutter path to adjust for the difference between actual and programmed cutter radii or diameters.
g50–g59 Adaptive Control	Reserved for adaptive control requirements.

g70 Inch Programming	Mode for programming in inch units. It is recommended that control turn on establish this mode of operation.
g71 Metric Programming	Mode for programming in metric units. This mode is cancelled by g70, m02, and m30.
g80 Axis Selection	Command that will discontinue any fixed cycle.

Fixed Cycle				*At Bottom*		
Number	*Code*	*Movement In*	*Dwell*	*Spindle*	*Movement Out To Feed Start*	*Typical Usage*
1	g81	Feed	—	—	Rapid	Drill, Spot Drill
2	g82	Feed	Yes	—	Rapid	Drill, Counterbore
3	g83	Intermittent	—	—	Rapid	Deep Hole
4	g84	Spindle Forward Feed	—	Rev.	Feed	Tap
5	g85	Feed	—	—	Feed	Bore
6	g86	Start Spindle, Feed	—	Stop	Rapid	Bore
7	g87	Start Spindle, Feed	—	Stop	Manual	Bore
8	g88	Start Spindle, Feed	Yes	Stop	Manual	Bore
9	g89	Feed	Yes	—	Feed	Bore

g90 Absolute Input	A control mode in which the data input is in the form of absolute dimensions.
g91 Incremental Input	A control made in which the data input is in the form of incremental data.
g92 Preload of Registers	Used to preload registers to desired values. No machine operation is initiated. Examples would include preload of axis position registers, spindle speed constraints, initial radius, etc. Information within this block shall conform to the character assignments of Table 1.
g93 Inverse Time Feedrate	The data following the feedrate address is equal to the reciprocal of the time in minutes to execute the blocks and is equivalent to the velocity of any axis divided by the corresponding programmed increment.

g94 Inches (Millimeters) Per Minute Feedrate	The feedrate code units are inches per minute or millimeters per minute.
g95 Inches (Millimeters) per Revolution	The feedrate code units are inches (millimeters) per revolution of the spindle.
g96 Constant Surface Speed Per Minute	The spindle speed code units are surface feet (meters) per minute and specify the tangential surface speed of the tool relative to the workpiece. The spindle speed is automatically controlled to maintain the programmed value.
m02 End of Program	A miscellaneous function indicating completion of workpiece. Stops spindle, coolant, and feed after completion of all commands in the block. Used to reset control and/or machine. Resetting control may include rewind of tape to the end of record character or progressing a loop tape through the splicing leader.*
m05 Spindle Off	Stop spindle in normal, most efficient manner; brake, if available, applied; coolant turned off.
m07–m08–m09 Coolant, On, Off	Mist (No. 2), flood (No. 1), tapping coolant or dust collector.*
m10–m11 Clamp, Unclamp	Can pertain to machine slides, workpiece, fixtures, spindle, etc.
m19 Oriented Spindle Stop	A miscellaneous function which causes the spindle to stop at a predetermined angular position.
m30 End of Tape	A miscellaneous function which stops spindle, coolant, and feed after completion of all commands in the block. Used to reset control and/or machine. Restting control will include rewind of tape to the end of record character, progressing a loop tape through the splicing leader, or transferring to a second tape reader.
m90–m99 Reserved for User	Miscellaneous function outputs which are reserved exclusively for the machine user.

Note: Additional commands may be required on specific machines. Unassigned code numbers should be used for these and specified on the Format Classification Sheet. On certain machines the functions described may not be completely applicable; deviations and interpretations should be clarified in the Format Classification Sheet. (See Appendix A.)

NATIONAL AEROSPACE STANDARD

4.3.9.3   Test Operation: (Continued)

4.   Press tape start button. The machine automatically removes drill, TL#123 from magazine and drills its own setup hole, returns to setup point and replaces drill with end mill, TL#789.

5.   The machine then automatically removes excess stock to the shape shown in the accompanying work sketch, leaving .040" stock for finishing, takes a finishing cut off the top surface, then returns to setup point and changes tools in spindle to the 1" diameter boring bar, TL#456.

6.   The machine automatically finish bores its own setup hole, returns to setup point, replaces TL#789- into spindle, and stops.

7.   Press start button. The machine now finish mills the shape as shown in work sketch.

As part of the finish cut three "friction-stiction" (low feed rate) tests are performed. See Page 19 for work sketch.

1.   As the cutter moves from right to left along surface (1), it pulls out .001"/inch. Upon completion of the piece, an indicator in the spindle nose may be run along surface (1) with zero motion. A comparison of indicator deflection and pull out demonstrates the ability of the machine to move an extremely small amount in the Z direction while moving a great amount in the X direction.

2.   As the cutter moves from bottom to top along surface (2), it moves to the right .001"/inch. For testing an indicator in the spindle nose may be run along surface (2) with zero X motion. As the cutter moves from left to right along surface (3), it moves down .001"/inch. For testing an indicator in the spindle nose may be run along surface (3) with zero Y motion.

Test piece tolerances and results:

Tolerance for 10" per Minute		Results	Tolerance for 5" per Minute
1. Angle Accuracies ±.003" overall			±.002" overall
2. Circle Accuracies ±.0015" on Radius			±.001" on Radius
	±.003" on Diameter		±.002" on Diameter
3. Depth Accuracies ±.0015"			±.001"

Friction-stiction test cuts shall not deviate from a uniformly rising straight path more than .0004" T.I.R.

4.3.9.4   Test Cuts

Maximum rated horsepower:

A straight cut 12" of length or more shall be made at the lowest spindle speed at which the machine is rated at full horsepower.

Cutter Type: Face Mill or shell end mill with tungsten carbide teeth.

Cutter Diameter: Suitable for the spindle speed that must be used in the test.

Material: Heat treated steel, 150,000 to 180,000 PSI (analysis optional).

4.4   Presentation of Inspection and Test Data

4.4.1   A certified inspection sheet listing the test results shall be presented to the customer prior to acceptance of the machine for shipment.

4.4.2   Data on the inspection sheet shall be in the same form as used in Section 4.3.

4.5   Acceptance of Machine:

4.5.1   Final acceptance will be at the customer's plant or designated point of installation after all requirements of this specification have been met. Acceptance for shipment will be based on compliance with this specification and satisfactory tests at the manufacturer's plant, with additional testing as required in Section 4.2.

CUSTODIAN:	Manufacturing Committee		
PROCUREMENT SPECIFICATION	TITLE		CLASSIFICATION Specification
None	NUMERICALLY CONTROLLED COMBINATION DRILLING, BORING, MILLING AND TAPPING MACHINES		**NAS** 978 Sheet 15

Copyright, 1965, Aerospace Industries Association of America, Inc. Reprinted by permission.

NATIONAL AEROSPACE STANDARD

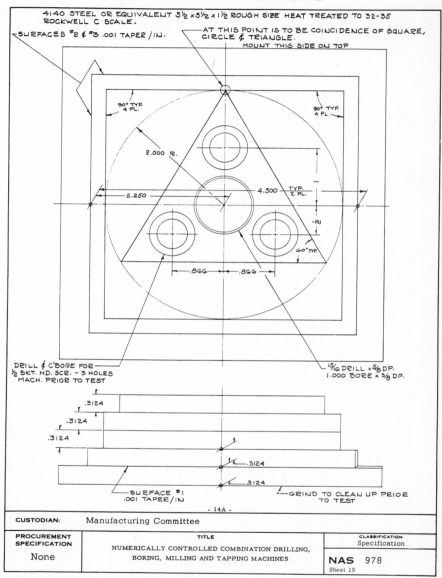

4140 STEEL OR EQUIVALENT 5½ x 5½ x 1½ ROUGH SIZE HEAT TREATED TO 32-35 ROCKWELL C SCALE.

SURFACES #2 & #3 .001 TAPER/IN.

AT THIS POINT IS TO BE COINCIDENCE OF SQUARE, CIRCLE & TRIANGLE.
MOUNT THIS SIDE ON TOP

90° TYP. 4 PL.

90° TYP. 4 PL.

2.000 R.

2.250

4.500   TYP. 2 PL.

60° TYP.

.866   .866

DRILL & C'BORE FOR ½ SKT. HD. SCR. - 3 HOLES MACH. PRIOR TO TEST

¹⁵⁄₁₆ DRILL x ⅝ DP. 1.000 BORE x ⅜ DP.

.3124

.3124

.3124

.3124

.3124

SURFACE #1 .001 TAPER/IN

GRIND TO CLEAN UP PRIOR TO TEST

- 14A -

CUSTODIAN:	Manufacturing Committee	
PROCUREMENT SPECIFICATION	**TITLE**	**CLASSIFICATION** Specification
None	NUMERICALLY CONTROLLED COMBINATION DRILLING, BORING, MILLING AND TAPPING MACHINES	**NAS** 978 Sheet 19

NATIONAL AEROSPACE STANDARD

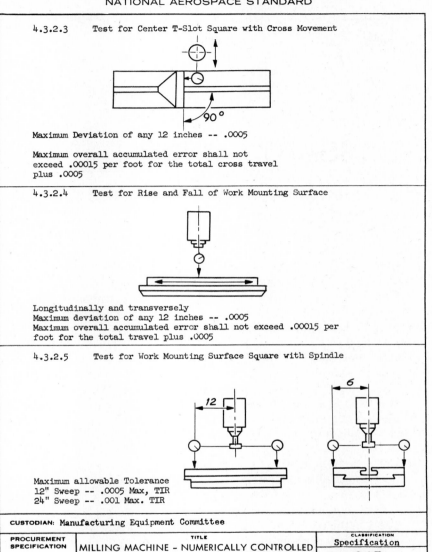

4.3.2.3   Test for Center T-Slot Square with Cross Movement

90°

Maximum Deviation of any 12 inches -- .0005

Maximum overall accumulated error shall not
exceed .00015 per foot for the total cross travel
plus .0005

4.3.2.4   Test for Rise and Fall of Work Mounting Surface

Longitudinally and transversely
Maximum deviation of any 12 inches -- .0005
Maximum overall accumulated error shall not exceed .00015 per
foot for the total travel plus .0005

4.3.2.5   Test for Work Mounting Surface Square with Spindle

12

6

Maximum allowable Tolerance
12" Sweep -- .0005 Max, TIR
24" Sweep -- .001 Max. TIR

CUSTODIAN: Manufacturing Equipment Committee

PROCUREMENT SPECIFICATION	TITLE	CLASSIFICATION
	MILLING MACHINE - NUMERICALLY CONTROLLED PROFILING AND CONTOURING	Specification
		NAS 913 Sheet 17

GLOSSARY

This glossary is based on *Automation Bulletin* 3B, published by the Electronic Industries Association, and used with their permission. This glossary does not include the entire *Bulletin*. A number of additions have been made, and some definitions have been reworded.

Absolute dimension (also called coordinate dimension) A dimension expressed with respect to the initial zero point of a coordinate axis. The zero point on some machines is established by the manufacturer at a fixed location, and on other machines it may be established by the N/C programmer.

AC or A/C adaptive control A system of sensors which changes speeds and/or feeds in response to the sensed pressure on the cutter.

AD-APT An ADaptation of the APT N/C programming language for use with computers with relatively small storage capacity.

Address A means of identifying information or a location in a control system. In numerical control the address is usually a letter. *Example:* The x in the command x 12345 is an address identifying the numbers 12345 as referring to a position on the X axis.

Aerospace Industries Association of America, Inc. (AIA) The national trade association of the manufacturers of aircraft, missiles, spacecraft propulsion, navigation and guidance systems, support equipment, accessories parts, and materials and components used in the construction, operation, and maintenance of these products.

Analog data The word "analog" comes from a Greek word meaning "proportional." Thus analog data usually comprise continuous information which varies in amount as the input signal (such as voltage or amperage) changes. The accuracy obtainable depends on the accuracy of the input and output measuring equipment. Common examples of analog data are the paper charts of temperature of a process from minute to minute and the variously shaped lines seen in an oscilloscope. Often contrasted to Digital data.

ANSI American National Standards Institute. Formerly the American Standards Institute, then the United States of America Standards Institute. This is a nonprofit organization which works with United States industry and the ISO to establish national and international standards as guides to aid the manufacturer, the user, and the general public.

APT An abbreviation of Automatic Programmed Tools. APT is a computer-based numerical control programming system which uses English-like symbolic descriptions of part and tool geometry and tool motion. It can be used only on large computers.

Arc clockwise An arc generated by the coordinated motion of two axes in which curvature of the path of the tool with respect to the workpiece is clockwise, when viewing the plane of motion in the negative direction of the perpendicular axis.

Arc counterclockwise An arc generated by the coordinated motion of two axes in which curvature of the path of the tool with respect to the workpiece is counterclockwise, when viewing the plane of motion in the negative direction of the perpendicular axis.

ASCII American Standard Code for Information Interchange.

AUTOMAP An abbreviation of AUTOmatic MAchine Programming. This is a subset of the APT program, for simple contouring work requiring only sloped lines and circular arcs. AUTOMAP can be run on relatively small computers.

AUTOSPOT An abbreviation of AUTOmatic System for POsitioning Tools. This is a computer program which uses an English-like language to describe the point locations and machine operations for point-to-point and straight-cut N/C machines. It can also program linear and circular interpolation.

AUXILIARY function A function of a machine other than the control of the coordinates of a workpiece or tool. Includes functions such as MISCELLANEOUS, FEED, SPEED, TOOL SELECTION, etc. Not a PREPARATORY function. Frequently the m function.

Backlash (ASA C85) A relative movement between interacting mechanical parts, resulting from looseness.

BCD (see Binary-coded-decimal system)

Binary code A code in which each allowable position has one of two possible states. A common symbolism for binary states is 0 and 1.

Binary-coded-decimal system (BCD) A number system which uses a binary-coded number to represent each digit in the decimal numbering system.

Bit An abbreviation of *bi*nary dig*it*. On the numerical control tape this is usually the presence or absence of a punched hole.

Block A word or group of words considered as a unit and separated from other such units by an END OF BLOCK character. This group of words provides information for a complete instruction to the N/C machine, such as a cutting location, a tool change, or a dwell instruction.

Buffer storage In N/C this refers to a place where information from a block of tape is electronically saved or stored so that it can be transferred to the MCU instantly. The next block is then put into the buffer storage while the previous command is being performed. This avoids waiting for the tape to go through the tape reader.

CAD Computer aided design. See Chap. 13.

CAM Computer aided manufacturing. See Chap. 13.

Canned cycle (see Fixed cycle)

CCW Counterclockwise. Rotating in a direction opposite to the normal motion of the hands of a clock.

Chad The pieces of material that are removed when punching holes in tape or cards.

Channel (see Track)

Character One of a set of letters, digits, or symbols which may be combined to express information. *Example:* The characters normally used in numerical control, those representing the decimal digits 0 to 9, the letters of the alphabet, and special characters such as TAB, END OF BLOCK, and END OF RECORD.

Circular interpolation A mode of contouring control available on some N/C machines which uses the information contained in a single block to produce an arc of a circle. The N/C machine does not require tangents or chords as in usual contour programs. The velocities of the axes used to generate this arc are varied by special circuits in the machine control unit.

Closed-loop system A system in which the output, or some result of the output, is fed back to the MCU for comparison with the input, for the purpose of reducing the difference to zero.

CMM Coordinate measuring machine, for quickly checking part dimensions.

CNC Computer numerical control. See Chap. 13.

Constant cutting speed The condition achieved by varying the speed of rotation of the workpiece relative to the tool inversely proportional to the distance of the tool from the center of rotation.

Contouring control system (NCC) A system in which the controlled path of the cutter can result from the coordinated, simultaneous motion of two or more axes. The axes do not necessarily move at the same rate.

Controller   (see Machine control unit)

Coordinate dimension   (see Absolute dimension)

CRT   Cathode ray tube. Used to display tape commands and operator instructions in conjunction with CNC or DNC.

Cutter compensation   An adjustment, normal to the cutter path, for the difference between the actual and the programmed cutter diameters. This difference may be due to cutter wear or use of a different-size cutter. This adjustment is often made manually on the operator's console.

Cutting speed   The rate, in feet per minute, or metres per minute, at which the work passes the cutting edge of the tool or at which the cutting edge of the tool passes through the workpiece. (*See* rpm.)

CW   Clockwise. Rotating in the same direction as the normal motion of the hands of a clock. *Note:* Frequently it is necessary to specify the viewing direction in order to avoid confusion in CW and CCW.

Delta dimension   (see Incremental dimension)

Digital data   Digits are the numbers 0 to 9, and digital data are expressed in numbers. The accuracy of digital data in N/C depends on the number of digits the electronic circuits are built to interpret and how accurately the machine can respond to the signals. All N/C programs in this book are written with digital data.

Director   (see Machine control unit)

DNC   Direct Numerical Control. Usually eliminates the tape and feeds the coded information directly from a large, distantly located computer.

Dwell   A timed delay of programmed or established duration.

EB or **EOB** (see END OF BLOCK signal)

Electronic Industries Association (EIA)   A national trade association representing manufacturers of electronic parts, equipment, and systems for consumer, industrial, and military applications. The subject of numerical control is one of many within the electronics area.

EL or **EOL**   END OF LINE (same as END OF BLOCK)

END OF BLOCK signal   A symbol, or indicator, that defines the end of one block of data. In the BCD notation, this is a single hole punched in track 8.

END OF PROGRAM   A MISCELLANEOUS function indicating completion of workpiece. Stops spindle, coolant, and feed after completion of all commands in the block. Used to reset control and/or machine. Resetting control may include rewind of tape or progressing a loop tape through the splicing leader. The choice for a particular case is specified by the N/C machine manufacturer.

END OF TAPE   A MISCELLANEOUS function which stops spindle, coolant, and feed after completion of all commands in the block. Used to reset control and/or machine. Resetting control will include rewinding of tape, progressing a loop tape through the splicing leader, or transferring to a second tape reader. The choice for a particular case is specified by the manufacturer.

FEED function   The relative velocity between the tool and the work due to motion of the programmed axis (axes).

FEED RATE NUMBER (FRN)   Used most often in contouring (NCC) systems, this is an f code which specifies a feed rate. The FRN is calculated according to formulas, which vary somewhat in different coding systems. It is **not** the inches-per-minute feed generally used with the f code in NPC machines.

FEED RATE OVERRIDE   A manual function directing the control system to modify the programmed feed rate by a selected percentage, which is often less than 100 percent.

FIXED BLOCK format   A format in which the number and sequence of words and characters appearing in successive blocks are constant. Note that the number of characters is always the same; thus all blocks are the same length.

Fixed cycle   A preset series of operations which direct machine axis movement and/or cause spindle operation to complete such action as boring, drilling, tapping, or combinations thereof.

FIXED SEQUENTIAL format   A means of identifying a word by its location in the block. Words must be presented in a specific order, and all possible words preceding the last desired word must be present in the block. Note that the **number** of characters may vary; thus the blocks are not necessarily the same length.

FLOATING ZERO (not to be confused with ZERO OFFSET)   A characteristic of a numerical machine control permitting the zero reference point on an axis to be established readily at any point in the travel. The control retains no information on the location of any previously established zeros.

Format   Physical arrangement of possible locations of holes or magnetized areas on the N/C tapes. Also, the general order in which information appears on the tape.

fpm   Feet per minute; same as sfpm (see Cutting speed)

FRN (see FEED RATE NUMBER)

Hold   An untimed delay in the program, terminated by an operator or interlock action.

IC   (see Integrated circuit)

Incremental dimension (also called delta dimension)   A dimension expressed with respect to the preceding point in a sequence of points.

In mathematics this would be called delta X, delta Y, etc., meaning the difference between the coordinate locations of two points.

Integrated circuit   A subminiature electric circuit made by a highly refined process involving chemicals and photography. Within approximately a 0.100 by 0.100 area the circuit may contain several transistors, resistors, capacitors, and inductances. These components can be seen and manufactured only by use of a low-power microscope. Reliability and speed of operation are exceptionally high.

ipm   Feed rate in inches per minute.

ipr   Feed rate in inches per revolution.

ipt   Feed rate in inches per tooth.

ISO   International Standards Organization. This is the group which establishes the international standards (SI). They have representatives from countries all around the world.

LETTER ADDRESS format (see WORD ADDRESS format)

Level (see Track)

Linear interpolation   A mode of contouring control which uses the information contained in a block to produce velocities proportioned to the distance moved in two or more axes simultaneously.

Machine control unit (MCU)   Sometimes referred to as the controller, or director. This is the brains of the numerical control system. It consists of a tape reader plus the electrical and electronic equipment needed to change the tape codes into machine commands. The MCU contains the memory, computational abilities, and relatively simple switching circuits as needed for simple or complex machines. The machine operator's controls and indicator lights may be mounted on this unit, or all or part of these controls may be contained in a separate console.

Manual data input   A means for the manual insertion of numerical control commands. May be by typewriter, special keyboard, or knurled wheels.

Manuscript (see Programming)   An ordered list of numerical control instructions.

MCU (see Machine control unit)

MDI (see Manual data input)

Memory   Equipment (usually electronic equipment) into which data can be stored and held for use at a later time.

MIS   Management Information System. Often a part of a CAM system.

MISCELLANEOUS function   An on-off function of a machine such as SPINDLE STOP, COOLANT ON, CLAMP.

National Aerospace Standards (NAS)   A series of AIA-developed standards covering such aerospace hardware as fasteners, fittings, and electrical items; specifications for aerospace materials, packaging materials, and machine tools; and testing procedures. Approximately one hundred standards were issued or revised in 1966. The year 1967 marked its twenty-fifth anniversary.

NCC Numerical contouring control (see Contouring control system).

NDT Nondestructive testing.

NPC Numerical positioning control (see Positioning control system).

OFF-line Used in computer operations to indicate that the work is not done by the computer, but by other, usually less expensive, equipment. For example, the computer may on-line-produce magnetic tape or punched cards, and these cards or tape are then converted to punched tape by a separate piece of equipment—a card-to-tape converter.

Open-loop system A control system that has no means for comparing the output with input for control purposes.

OPTIONAL STOP A MISCELLANEOUS function command similar to a PROGRAM STOP, except that the control ignores the command unless the operator has previously pushed a button to recognize the command.

Parity bit In tape control this is (in the BCD code) the hole in track 5 which is used when necessary to make all BCD coding include an odd number of holes. A checking device to help in determining if the tape punch and tape reader are operating correctly.

Part programmer The person who writes the manuscript of the hand or computer program for machining a workpiece according to specifications, on a numerical control machine.

Part programming, computer The preparation of a manuscript in computer language and format required to accomplish a given task. The necessary calculations are to be performed by the computer.

Part programming, hand or manual The preparation of a manuscript in machine control language and format required to accomplish a given task. The necessary calculations are to be performed manually.

PLANNED STOP (see OPTIONAL STOP)

Point-to-point control system (see Positioning control system)

Positioning control system (NPC) A system in which the controlled motion is required only to reach a given end point, with no path control during the movement from one end point to the next.

Postprocessor A set of computer instructions which transform tool centerline data into machine motion commands using the proper tape code and format required by a specific machine control system. Instructions such as feed rate calculations, spindle-speed calculations, and auxiliary function commands may be included.

PREPARATORY function A command changing the mode of operation of the control, such as from positioning to contouring, or calling for a fixed cycle of the machine.

Programmer (see Part programmer)

Programming   The ordered listing of a sequence of events designed to accomplish a given task.

PROGRAM STOP   A MISCELLANEOUS function command to stop the spindle, coolant, and feed after completion of other commands in the block. It is necessary for the operator to push a button in order to continue with the remainder of the program.

Prototype   A first, or original, model which, after testing and refining, is the basis for the final product.

Readout, command   Display of absolute position of the cutter, as called for by the tape command. Does not change as the cutter moves to the position. *Note:* In many systems the readout information may be taken directly from the dimension command storage. In others it may result from the summation of command departures.

Readout, position   Display of the absolute position of the centerline or cutting end of the cutter as it moves along its path. Readout changes as the cutter position changes.

Resolution   The shortest distance between two adjacent cutter locations which can be distinguished one from the other by the numerical control system. Commonly available resolutions are 0.02 or 0.005 or 0.002 mm [0.001 or 0.0002 or 0.0001 in.].

Retrofit   In numerical control this refers to the fitting of numerical control equipment to a machine which was originally operated by manual or tracer control. The machine may require considerable alteration and repair so that the N/C controls will function properly and accurately.

Row   A path perpendicular to the edge of a tape along which information may be stored by the presence or absence of holes or magnetized areas.

rpm   Revolutions per minute, or "speed" in an N/C program.

SEQUENCE NUMBER   A number identifying each block of information on a tape. This number is assigned by the part programmer or by the computer.

SEQUENCE NUMBER READOUT   Display of the SEQUENCE NUMBER which was punched on the tape.

sfpm   Surface feet per minute, or cutting speed. Same as fpm.

SI   Systéme Internationale. The internationally agreed on metric system of units.

Spindle speed   The rate of rotation of the machine spindle, usually expressed in rpm. Usually, the S function in N/C programs.

Straight-cut control system   A system in which the controlled cutting action occurs only along a path parallel to linear or circular machine ways. Frequently combined with point-to-point capabilities.

TAB SEQUENTIAL **format**   A means of identifying a word by the number of tab characters preceding the word in the block. The first character in each word is a tab character. Words must be presented in a specific order, but all characters in a word, except the tab character, may be omitted when the command represented by that word is not desired.

TOOL **function**   A command identifying a tool and calling for its selection either automatically or manually. The actual changing of the tool may be accomplished by a separate tool change command.

Tool offset   A correction for tool position parallel to a controlled axis.

Track (formerly channel, or level)   A path parallel to the edge of a tape along which information may be stored by the presence or absence of holes or magnetized areas.

Transducer   A device for changing one form of energy into another form, for example, from an electric signal into mechanical motion, as in a relay.

United States of America Standards Institute (see ANSI)

VARIABLE BLOCK **format**   A format which allows the number of words in successive blocks to vary. This applies to both TAB SEQUENTIAL and WORD ADDRESS programs. See FIXED BLOCK format.

Verify   To check, usually automatically, one typing or recording of data against another in order to minimize human and machine errors in the punching of tape or cards.

Word   A set of characters which gives a single complete instruction to the N/C machine. For example, X06450 and m06 are both "words."

WORD ADDRESS **format**   Addressing each word of a block by one or more characters which identify the meaning of the word.

ZERO OFFSET (not to be confused with FLOATING ZERO)   A characteristic of a numerical machine control permitting the fixed zero point on an axis to be shifted readily over a specified range. The control retains information on the location of the "permanent" zero. The shift available may be only ± 0.12 mm or may be FULL ZERO SHIFT, which will operate to any place on the machine table.

ZERO SHIFT (see ZERO OFFSET)

INDEX